INTERNATIONAL SERIES OF MONOGRAPHS IN

NATURAL PHILOSOPHY

General Editor: D. ter Haar

Volume 58

WAVE MECHANICS
AND ITS APPLICATIONS

WAVE MECHANICS
AND ITS APPLICATIONS

BY

P. GOMBÁS

AND

D. KISDI

BUDAPEST
HUNGARY

PERGAMON PRESS

OXFORD NEW YORK TORONTO
SYDNEY BRAUNSCHWEIG

Pergamon Press Ltd., Headington Hill Hall, Oxford
Pergamon Press Inc., Maxwell House, Fairview Park, Elmsford,
New York 10523
Pergamon of Canada Ltd., 207 Queen's Quay West, Toronto 1
Pergamon Press. (Aust.) Pty. Ltd., 19a Boundary Street,
Ruschcutters Bay, N.S.W. 2011, Australia
Vieweg & Sohn GmbH, Burgplatz 1, Braunschweig

The original *Bevezetés a hullámmechanikába és alkalmazásaiba* was published by
Akadémiai Kiadó, Budapest

Translated by J. Schanda

Co-edition of Pergamon Press, Oxford and Akadémiai Kiadó, Budapest

Library of Congress Cataloging in Publication Data

Gombás, Pál, 1909–
  Wave mechanics and its applications.

  (International series of monographs in natural
philosophy, v. 58)
  Translation of Bevezetés a hullámmechanikába és
alkalmazásaiba.
  1. Wave mechnics. I. Kisdi, Dávid, joint author.
II. Title.
QC174.2.G5913   530.1'24   73-5789
ISBN 0-08-016979-1

First English edition 1973

PRINTED IN HUNGARY

Contents

CONTENTS

Preface

WITH our book we intend to offer an introduction to the quantum mechanical theory of the atom and its applications. Our aim is to arouse interest in nuclear physics and we hope that as many young scientists as possible will join in the theoretical research work of the atom going on all over the world. The book can be divided into two parts. The first part deals with the fundamentals of quantum mechanics in a rather rudimentary way selecting from the vast material mainly those parts which prepare and establish the second part, the applications. This first part of the book, as well as Sections 30, 31, 32, 37, 44 and 46 of Chapter 4 correspond, apart from slight alterations, to the first three chapters of another book by one of the authors (P. Gombás, *Introduction to Atomic Theory*, Institute for Continuation Course for Engineers, Budapest, 1946). This part deals with the detailed solution of a great many definite problems of wave mechanics. The aim of a detailed discussion of the problems is to make young scientists acquainted with concrete quantum mechanical calculations; by going in detail through these calculations they should be able to apply their newly acquired techniques. This is a fundamental and rather difficult part of the training of young researchers, and that is why special attention was paid to this in our book.

It is to be regretted that among the young generation we may encounter the view that "real" and "modern" theoretical physics is only that one that reaches the final result without calculations, and, for example, they are not interested in the determination of lattice parameters or of energy band structures of a metal from first principles, only because it includes numerical calculations as well. They regard this as uninteresting and out of date. We cannot but disagree with this idea.

On the other hand, it is absolutely true that the exploration, the recognition or the deduction of the general theoretical physical relations is the most beautiful part of the theoretical physical work, one might say, its crown. But this success falls only to a very few people's lot. And the way to it leads, in our opinion, not by neglecting the sometimes tedious work of calculations but just through it. We know, of course, that there are exceptions to this among the greatest physicists, but it is also true that the majority of just the greatest physicists began their career with the solution of simpler problems requiring detailed calculations.

The present book, about which the Table of Contents gives more detailed information, is divided into six chapters. Chapter 1 discusses the most important experimental basis of quantum theory. Chapter 2 sums up the old, so-called Bohr atomic theory which, although an outworn concept, is indispensable for understanding the modern theory. Chapter 3 speaks about the essential basis of wave mechanics. In Chapter 4 we discuss the solution of the Schrödinger equation in those simple cases when the solution can be given exactly. Chapter 5 surveys the quantum theory of scattering in a potential field and discusses the theory of scattering resonances in detail. Finally, Chapter 6 speaks about the three, very simple approximate methods of wave mechanics: perturbation theory, the variational method and the Born approximation in scattering theory.

The application of these approximate methods is demonstrated by problems which have already become classical in nuclear physics.

The authors are highly obliged for their help in editing this book to Mrs. Gy. Huszár, Mrs. I. Kocsis, Miss O. Kunvári, Miss Zs. Ozoróczy, Mrs. S. Rajos, and Mrs. I. Tamás, co-workers of the research group for Theoretical Physics of the Hungarian Academy of Sciences.

Introduction

ATOMIC theory taken in a wider sense reaches back to the Greeks where, however, it did not rely on experimental observations and results became of speculative character, so that this has only historical importance today. It was only in modern times, at the time of the great development in natural sciences, at the end of the seventeenth century, that atomic theory in a narrower sense began to take shape, by the help of which it became possible to explain certain natural laws. The first major result was the interpretation of gas laws on the basis of a corpuscular theory. The corpuscular theory was adopted in other fields as well; for instance, Cauchy tried to explain the characteristics of elastic bodies on the basis of this theory. One of the main supports of this theory was offered by Dalton's law, according to which elements may form compounds only in definite mass ratios. By the discovery of the electron and H.A. Lorentz's theory of electrons attributing the electromagnetic behaviour of bodies to the charges moving within them, the theory got a significant stimulus at the end of the last century.

Within the past fifty years the development of the theory has been extremely rapid, discovery followed discovery, partly through experimental results, and partly through fruitful theories. In 1900 Planck, with his important investigations referring to the blackbody radiation, laid the foundation of a highly significant theory, that of the quantum theory, a theory which has become the starting-point for modern nuclear theory. When in 1913 Bohr created his nuclear theory, it meant a new phase in the development. Bohr's theory proved to be very fertile and it became the starting point of the monumental theory of atomic and molecular spectra which spectra provide an extraordinarily rich experimental material for the understanding of the properties of the atom and the molecule, respectively.

In 1925 the development of the theory entered a significantly new phase. It was namely then that Heisenberg created matrix mechanics which was followed within only a few months by the wave mechanics of Schrödinger. Matrix mechanics and wave mechanics started from entirely new concepts and deepened the quantum theory considerably, they even transformed in many respects our whole concept of the world. Formally matrix mechanics and wave mechanics differ significantly from each other, one operates with matrices, the other with differential equations. But it turned out soon that these two theories so different in their appearance are fundamentally the same: they are two different representations of quantum mechanics. Quantum theory took an extremely rapid new lease of life, so that today after four decades it rivals the old classical theories as to comprehensiveness and completeness.

Quantum mechanics can interpret at least qualitatively the phenomena and processes taking place in the electron shell, and in most cases it can even account for them quantitatively as well. But quantum mechanics has outgrown the frame of nuclear physics and its importance proved to be significant in almost every branch of science. The quantum mechanical examination of molecules has prepared the way for a new branch of science: quantum chemistry. Quantum mechanics has achieved crucial success in the physics of solids and liquids, and one of its greatest results here is the theoretical explanation of superconductivity and superfluidity.

A new branch of science, quantum biology, examines the structure and transformations of molecules which are basically important from the point of view of life processes. Quantum mechanical effects occurring in the inner structure of the nucleus are discussed by nuclear physics, whereas quantum electrodynamics deals with quantum processes of the electromagnetic phenomena. The part of quantum mechanics discussing the transformations of elementary particles is the quantum field theory. In the above-mentioned fields, however, but especially in the physics of elementary particles no theory has been developed yet which could be regarded as final.

PART ONE

CHAPTER 1

Experimental Basis

1. The Electron

Electrons, the elementary particles of electricity, may be studied in connection with several phenomena, the best known of which are the cathode rays. Cathode rays start from the cathode (from the negative electrode); in the case of electrical discharges produced in highly evacuated tubes, these rays travel along a straight line and produce a greenish-blue luminescence on the glass envelope of the tube.

They are curved in an electric or magnetic field from which it can be ascertained that they carry a negative charge. All the properties of the rays can be explained with the assumption that they consist of elementary particles of negative charge, i.e. of electrons. From the deviation of the rays in electric and magnetic fields the ratio $-e/m_0$ of the electron charge, $-e$, to the electron mass, m_0, could be determined, as well as the electron velocity, v. It was found that the velocity of the electrons, which depends on the potential difference between the cathode and the anode (the positive electrode of the discharge tube) changes within broad limits and at high voltages it will approximate the velocity of light (3×10^{10} cm/sec). In case of low and moderate velocities the value of e/m_0 was found to be

$$\frac{e}{m_0} = 5.273 \times 10^{17} \text{ el. stat. units}. \tag{1.1}$$

We have to mention that this ratio depends on the velocity which, however, plays a role only in case of very high velocities (because, according to the theory of relativity, m_0 depends on the velocity). During experiments it was learnt that the absolute value of the charge of the electron "e", is the smallest occurring elementary particle of electricity. For its measurement in 1909 Millikan and Ehrenhaft worked out, independently from each other, the following method. They introduced vaporized oil or mercury drops or metal particles between the two horizontal metal plates of a condenser, where a vertical homogeneous electric field can be produced. These particles either possess an electric charge because of the way of the preparation itself or can be provided with a charge by suitable means, e.g. by means of a photoelectric effect or by ionizing the air when they can take ions from it.

Without an electric field the particles fall downwards with a constant velocity because of the joint impact of the force of gravity and of the frictional force on the particles and this can be followed and measured by a microscope. If in this state the velocity of a spherical particle with radius a is v_0, according to Stokes' theory the frictional force having an effect on it is $6\pi\eta a v_0$, where η is the viscosity coefficient of air. The weight of the particles keeps balance with this force which, in air, is $(4\pi a^3/3)(\rho-\sigma)g$, where ρ is the density of the particle, and σ is that of the air. Thus the following relation is valid

$$\frac{4\pi a^3}{3}(\rho - \sigma)g = 6\pi\eta a v_0. \tag{1.2}$$

From this a, the radius of the spherical particle can be calculated. If we produce between the plates of the condenser an electric field with a field strength E, the falling velocity of the drop will change to a value v_1, it will be lower or higher than the previous one depending on the sign of its charge. If we denote the charge of the particle by e_i, we obtain in this case

$$\frac{4\pi a^3}{3}(\rho - \sigma)g + Ee_i = 6\pi\eta a v_1, \tag{1.3}$$

namely the electric force Ee_i, having an effect on the particle, adds also to the weight of it. From these two equations the charge e_i can be calculated

$$e_i = \frac{6\pi\eta a(v_1 - v_0)}{E}. \tag{1.4}$$

The result of the Millikan experiments is that the charge of the single particles is always a low integer multiple of a certain definite elementary charge e. The unit of this elementary charge is

$$e = 4.803 \times 10^{-10} \text{ el. stat. units.} \tag{1.5}$$

It is important that we have to deal with small integer multiples, as this is a precondition for the existence of an elementary charge. Large integer multiples do not prove anything, due to the effect of measurement errors.

With the help of (1.1) and (1.5) the mass m_0 of the electron can be calculated:

$$m_0 = 9.108 \times 10^{-28} \text{ g.} \tag{1.6}$$

Thus the mass of the electron is extremely small, it is 1850 times smaller than the mass of the lightest atom, the H atom.

Radioactive substances emit three different radiations: α-, β- and γ-rays. The α-rays consist of positively charged He atoms, i.e. He ions, with velocity roughly 1/15–1/20 of the velocity of light. The penetration depth of these rays is small, even the most penetrative ones are totally absorbed by a 9-cm-thick layer of air. The β-rays are essentially the same as the cathode rays, only the initial velocity of the electrons is very high, approaching the velocity of light. Their penetration depth is also higher than that of the α-rays. It has to be mentioned that the γ-rays are essentially different from the α- and β-rays. While the α- and β-rays consist of electrically charged particles, the γ-rays are a form of electromagnetic radiation, and do not carry a charge.

In the case of the photoelectric effect, sometimes called after its discoverer Hallwachs effect, rays can be observed, the behaviour of which is essentially the same as that of the cathode and β-rays. The essentials of the photoelectric effect are the following: due to the irradiation of metals with UV- or X-rays—in some cases even with light irradiation—the metals emit a negatively charged radiation, and as a consequence of this the metal plate becomes positively charged. By the investigation of the bending of these rays in electric and magnetic fields it was established that they consist of electrons; they are essentially cathode rays, but the velocity of these electrons is much smaller, in the order of 0.001 times the light velocity.

It has to be mentioned that glowing bodies emit electrons too, the velocity of which is roughly equal to that of the photoelectrons.

In connection with the above-mentioned effects we got acquainted with one of the most important building blocks of the atom, namely with the electron, the atom of electricity, the charge of which is the so-called elementary charge. According to our present knowledge smaller charges do not exist. At the same time the methods of separating electrons from some materials have been demonstrated, too. The investigation of these effects, and especially of the cathode rays, has transformed considerably the atom concept of the last century, according to which atoms are the smallest indivisible parts of the elements.

In the light of the knowledge gained by the investigation of cathode rays, this concept became untenable, as it turned out that electrons can be separated from the atoms. Thus atoms have to contain electrons. At the same time the atoms cannot consist only of electrons, first of all because such a formation would explode due to the repulsion between the electrons, and secondly because in this case atoms would be negatively charged, in contradiction to experience, according to which atoms are neutral. Thus a positive charge has also to be somewhere in the atom too, and this positive charge neutral-izes the charge of the electrons.

2. The Scattering of α-particles. Rutherford's Atom Model

Important information on the positive charge existing in the atom has been gained from the basic experiments performed by Rutherford, Geiger, Marsden and Chadwick dealing with the scattering of α-particles as they penetrate a thin metal foil. α-particles are, as mentioned earlier, positively charged He atoms. Their mass is roughly 7400 times higher than the electron mass. If atoms are bombarded by α-rays, important information can be gained about the structure of the atoms. Such a heavy particle will deviate from its original path only due to a strong effect – electrons, for instance, will not deviate it from its original path. Thus it can be hoped that by such experiments information can be gained about the positive constituents of the atom. The essentials of these experiments are the following: If an α-ray falls on a ZnS screen, it produces fluorescence; if the ray is weak enough, under high magnification it can be seen that the fluorescence consists of single flashes, scintillations, produced by single impinging α-particles. This gives a possibility for counting the α-particles. Rutherford placed into the way of the α-ray a very thin gold foil, so that the rays could still penetrate it. Without the gold plate only a very small area of the ZnS screen lit up, where the α-particles travelling in a straight line reached the screen. With the gold foil in the path of the radiation the luminescing spot became broader, thus a part of the α-particles slightly deviated from its original direction, but α-particles could be observed in all other directions too, some of them deviated from the original direction to a higher extent; it was even possible to find α-particles, reflected from the gold plate and travelling in an opposite direction – their number was, however, very small. Generally, as was con-cluded from the very tedious counting of the scintillations, the number of deviated particles dropped sharply with increasing angle of deviation: most of the particles penetrated the foil without any change of direction. From the results of these experiments Rutherford came to the very important conclusion that the positive charge and almost the whole mass of the atom is concentrated in a very small volume, in the so-called nucleus, the dimensions of which are of the order of 10^{-12}–10^{-13} cm. The electrons are about 10^{-8} cm from the nucleus, they revolve around it like planets, and their mass is negligible compared with that of the nucleus.

By the help of this simple nuclear model the scattering of α-particles can be treated theoretically, but at the present stage we cannot go into details. The scattering is the result of a Coulomb-like electrostatic repulsion between the nucleus and the α-particle. Supposing that every α-particle will be scattered only once during the penetration through the foil, the following scattering formula can be deduced:

$$\Delta n = n \frac{4\pi s N Z^2 e^2 \cos \vartheta/2}{m^2 v_0^4 \sin^3 \vartheta/2} \Delta \vartheta . \qquad (2.1)$$

This gives the number of particles in a direction between ϑ and $\vartheta + \Delta \vartheta$ from the original direction, if the number of impinging particles is n and the thickness of the foil is s, N is the number of atoms per cm^3 of the scattering material (that is to say the number of scattering nuclei), Z is the number of elementary charges in the nucleus, e is the elementary charge, m the mass of the α-particle, and v_0 its starting velocity.

This scattering formula has been established experimentally for heavy nuclei. Thus the principal suppositions used in the deduction of the formula were proved.

By comparing the scattering of different materials a very important result was gained: Z, the nuclear charge in units e is equal to the serial number of the element, i.e. to that number which shows the place of the element in the periodic system. The elements have been arranged mainly according to their atomic weight, but it turned out that the order was complete only then if the sequence of some elements had been inverted. There is a reversal in the atomic-weight order of sequence at the elements cobalt–nickel, argon–potassium, tellurium–iodine. It was a significant recognition that not the atomic weight but the nuclear charge is of primary importance in arranging the atoms into a periodic table. This governs also the physical and chemical behaviour of the elements. The nuclear charge is increasing throughout the periodic system without any anomaly such as found in the case of the atomic weight. As a neutral atom has no charge, the nucleus is surrounded by just as many electrons as are necessary for the neutralization of the positive charge of the nucleus. Thus the following important relation holds: the nuclear charge in units e is equal to the serial number and is equal to the number of electrons surrounding the nucleus. Atoms which possess an electric charge are called ions. These are the atoms which have lost one or more electrons, and thus the electrons cannot compensate the positive charge of the nucleus completely. It can happen that the atom binds more electrons than are needed to compensate the nuclear charge, in such cases one speaks of a negative ion.

Generally the atomic weight increases with the serial number (apart from the afore-mentioned exceptions), but usually much more quickly than the serial number. The serial number can be determined very accurately by means of X-ray investigations on the basis of the Moseley law.

The investigation of the scattering of α-particles gives a possibility of investigating whether the Coulomb forces are valid in the very proximity of the nucleus or not. The scattering formula has been deduced by supposing the validity of the Coulomb forces, thus if the Coulomb forces are valid then – according to formula (2.1) – $\Delta n v_0^4$ is, for a given element in a given direction, constant. By changing v_0 the validity of this expression can be checked. It has been established that the Coulomb forces are valid up to a distance of 10^{-12}–10^{-13} cm. Thus the nuclear diameter has to lie in this order of magnitude. The atomic diameters are of the order of 10^{-8} cm, thus they are 10,000–100,000 times larger. In this space with a radius of 10^{-8} cm are the electrons. It has to be mentioned that from other reasoning it turns out that the electrons themselves have a size of about 10^{-13} cm.

3. The Franck–Hertz Experiment

Further basic knowledge can be gained about the structure of the atoms by investigating the scattering of electrons by atoms as was done by J. Franck and G. Hertz in their experiment.

Their first important observation was as follows: if the velocity of the electrons that collide with the atoms of a gas does not reach a critical value typical of the given gas, the collision can be regarded as elastic. This can be demonstrated in the following manner (Fig. 1). The electrons emitted by the incandescent filament i and accelerated by a potential difference between i and the mesh h reach an average velocity depending on this potential difference. The electrons collide with the gas atoms in the space between i and h, and penetrate the mesh. The metal plate f is placed near the mesh, and there is a counter-potential applied between h and f, which is, however, smaller than the accelerating potential. By varying this reverse potential the velocity distribution of the electrons reaching f can be determined. From the current flowing between i and f the number of electrons reaching f can be calculated, from the reverse potential between f and h the velocity of the electrons at the mesh (h) can be determined, as only those electrons reach f whose velocity is high enough to overcome the counter-potential.

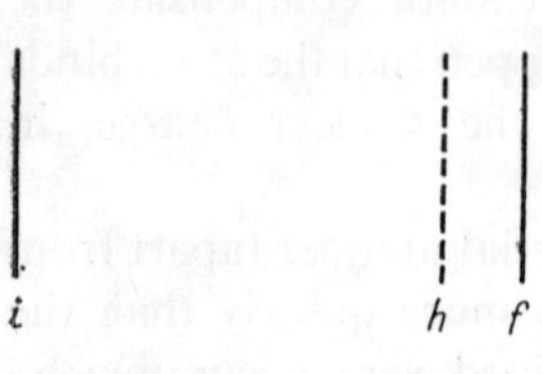

FIG. 1. Sketch of the Franck–Hertz tube

Between i and h the electrons suffer a great number of collisions with the atoms, the number of which varies with the distance between i and h. If the end-velocity of the electrons is independent of the number of collisions, the electron atom collision is elastic. From this Franck and Hertz concluded that apart from those gases which bind electrons, i.e. from the electron-affine

8

gases, one may always find a velocity interval where the collisions are elastic. The collision of slow electrons with gas atoms is always elastic.

Now the following experiment can be performed. The potential difference between f and h (the counter-potential) should be kept constant, at the same time the potential (V) between i and h should be increased continuously. At low V values the collisions will be found to be elastic, but by increasing V one reaches a value where they stop being elastic. Measuring the current between i and f one finds that in the region where the collisions are elastic the current grows with V, because the velocity of the electrons increases with V and thus also the number of electrons reaching f per unit time, i.e. the current. As the critical V where the collisions stop being elastic is reached, the electrons lose their velocity during the collision, and thus the current drops sharply. If the voltage V is increased further, the place where the electrons reach the critical velocity of inelastic collision will be not immediately before h, but nearer to i, thus the electrons will gain velocity between this place and h, due to the accelerating field. If the voltage is increased further, electrons will gain enough energy for a further inelastic collision before reaching h. By a further increase in V this process will repeat itself. Thus the current (I) as a function of V will show equidistant maxima and minima. By the help of this method the so-called first critical potential can be determined. Electrons having reached this potential will collide with the gas atoms inelastically (Fig. 2).

Using this method it was not possible to increase the velocity of the electrons above the value related to the first critical potential. Modifying slightly the experimental setup this can be achieved. Two screens have to be applied between i and f; these are h_1 and h_2, where h_1 is placed so close to i that in the low-pressure atmosphere of the gas the electrons practically cannot collide with the atoms. The potential difference between i and h_1 accelerates the electrons. Then the elec-

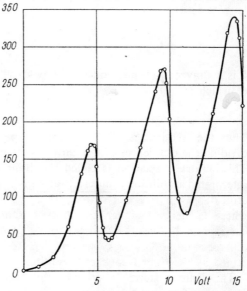

FIG. 2. Current versus voltage characteristics in the Franck–Hertz tube

trons reach the space between h_1 and h_2 where there is no electric field, but the electrons collide with the atoms. Between h_2 and f there is again a counter-potential. This method enables us to bring electrons with a chosen velocity into the space between h_1 and h_2, where the collisions occur. In the experiment the current between i and f is investigated which shows a minimum whenever the electrons suffer an inelastic collision. With the help of this it became possible to determine a whole series of critical potentials. Increasing the accelerating potential, i.e. the velocity of the electrons, continuously, at a given value one reaches the ionization of the atoms (an electron is separated from the atom, the atom separates into a positive ion and a negative electron).

In mercury vapour, for instance, the first critical potential is 4.9 V, the second 6.7 V and ionization takes place at 10.2 V.

The critical potentials, and the energy values corresponding to them, belong to the most important constants of the atoms and molecules. At the first inelastic collision the gas begins to emit light of a well-defined wavelength. There exists a basic relation between the frequency of these spectral lines and the energy values of the critical potentials: If $E_1, E_2, \ldots, E_i, \ldots$ are the energies corresponding to the critical potentials (these are the energies of the electron gains as it travels through a potential difference equal to the critical potential, for example, for Hg: 4.9 eV; 6.7 eV, etc.) and v_{ik} denotes the frequencies of the emitted light, the following relation holds:

$$hv_{ik} = E_i - E_k ; \quad v_{ik} = \frac{E_i - E_k}{h} ; \tag{3.1}$$

h is a universal atomic constant, the so-called Planck constant, the value of which is

$$h = 6.626 \times 10^{-27} \text{ erg} . \tag{3.2}$$

Thus the frequencies of the occurring spectral lines can be obtained by dividing the differences of the corresponding critical electron energies by h. This important relation is called the Bohr frequency condition. It can be interpreted as follows: each electron transfers the energy corresponding to a critical potential to a gas atom, thus the atom gets into an energy state where its energy is higher than in the normal state by the value of the critical energy. If the atom returns to its ground state it emits the energy difference in the form of radiation. The frequency of the radiation is determined by the Bohr frequency condition.

The Franck–Hertz experiment thus showed the surprising and very important fact that the atoms do not absorb and emit light *ad libitum* but only well defined quantities of energy.

4. The Magnetic Moment of Atoms

The existence of magnetic moments of atoms was first demonstrated by Stern and Gerlach in 1921. The basic idea of their method by means of which they determined the magnetic moment of silver atoms is as follows. A homogeneous magnetic field aligns the atoms having a magnetic moment, but does not produce a translation. Translational forces occur only in inhomogeneous fields. This can be realized if one substitutes the moment by a dipole, irrespective of whether it is perhaps produced by a system of currents. This is possible as a system of currents also corresponds to a dipole.

Let us consider a small magnet (length l), the pole strengths of which are $+\gamma$ and $-\gamma$. The absolute value of the moment of this magnet is

$$\mu = l\gamma. \tag{4.1}$$

A homogeneous magnetic field with a field strength $\mathfrak{H}$ produces a torque on this dipole the absolute value of which is (see Fig. 3).

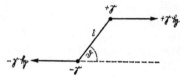

$$M = -\mu\mathfrak{H} \sin \vartheta, \tag{4.2}$$

ϑ is the angle between μ and $\mathfrak{H}$. The negative sign means that the field tries to make the angle smaller. In this case no translational forces occur.

Fig. 3. Magnetic dipole in a homogeneous magnetic field

If the field is inhomogeneous the magnetic forces at the two ends of the dipole are not equal any longer. If the field strength at the $+\gamma$ pole is $\mathfrak{H}'$ and at the $-\gamma$ pole it is $\mathfrak{H}$, the translational force is as follows (for the sake of simplicity it has been supposed that the field inhomogeneity occurs only in the x-direction):

$$\varDelta\mathbf{K} = \gamma\mathfrak{H}' - \gamma\mathfrak{H} = \gamma\left(\mathfrak{H} + \frac{\partial\mathfrak{H}}{\partial x}\varDelta x\right) - \gamma\mathfrak{H} = \gamma\left(\mathfrak{H} + \frac{\partial\mathfrak{H}}{\partial x}l\cos\vartheta\right) - \gamma\mathfrak{H},$$
$$\tag{4.3}$$

i.e.
$$\varDelta\mathbf{K} = \mu\cos\vartheta\,\frac{\partial\mathfrak{H}}{\partial x}. \tag{4.4}$$

Thus the force acting in the x-direction is equal to the component of the moment in the x-direction times the gradient of the field in the same direction.

11

The experimental apparatus of Stern and Gerlach was as follows (Fig. 4). They heated the metal (K) under investigation in an evacuated glass tube, thus producing atom-rays in every direction (i.e. the vapour of the metal evaporates away). The velocity of the particles is determined by the temperature. By placing two diaphragms into the way of the atom-beam they succeeded in producing a ray of given direction. This beam was led through an inhomogeneous magnetic field, which was produced by taking as the poles of the magnet on one side a flat profile, on the other side a wedge-shaped one. The atomic beam was then gathered by a plate where the condensation of the atoms left a mark.

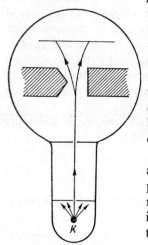

FIG. 4. Basic sketch of the Stern–Gerlach experiment

Stern and Gerlach used in their first experiment a silver beam. This experiment led to a very important conclusion: if the silver atoms possess a moment, the translation force produced by the inhomogeneous field will deviate the atoms and thus the beam. If the moment of the atoms has a given value, but the direction of it is at random, a widening of the beam may be expected. Atoms whose moment is either in the direction of the field or opposite to it will be found at the two ends of this widened patch. As a matter of fact, the experiment showed a different pattern: Stern and Gerlach found two patches on the plate (a double line), that is the beam parted into two, demonstrating that the magnetic moment of the atoms has a definite direction if placed in a magnetic field, one part of them is lined up parallel to the field while the other part is antiparallel.

The magnitude of the moment could be determined as well, and they obtained the following value:

$$\mu = 9.2 \times 10^{-21} \ \text{G/cm}^3. \tag{4.5}$$

The numerical value of the moment was, within the limits of experimental error, equal to one Bohr magneton. This can be interpreted on the basis of the atom theory, where the Bohr magneton is the elementary unit of the magnetic moment. This will be discussed in detail in the chapter dealing with the Bohr theory.

The numerical value of the Bohr magneton, as derived from the atomic theory, is

$$\mu_B = \frac{eh}{4\pi m_0 c} = 9.283 \times 10^{-21} \ \text{G/cm}^3, \tag{4.6}$$

where h is Planck's constant, e the elementary charge, m_0 the electron mass.

The result of the above-mentioned experiment is surprising and it discloses a very important atomic feature: the existence of an elementary magnetic moment and the fact that this moment lines up in a magnetic field only in given directions.

Other observations show that the magnetic moments of the atoms are always some integer multiples of the elementary magnetic moment. The magnetization of the bodies can be regarded as the alignment of the magnetic moments of the atoms. This mechanism is predominantly responsible for the occurrence of magnetization.

The magnetic moment is coupled with an angular momentum, as shown by the Einstein–De Haas experiment. A long cylindric soft-iron rod is suspended by an elastic thread in such a manner that it can rotate around its longitudinal axis and it is placed into a coaxial solenoid. If current is led into the solenoid magnetization occurs in the soft-iron; if the current is reversed the soft-iron remagnetizes and at the same time it rotates around its axis, that is, a torque is produced. As the magnetic moment in the soft-iron rod changes to the opposite direction the angular momentum changes too. This shows that the angular momentum is associated with the magnetic moment. By measuring the torque the angular momentum associated with a given magnetic moment can be determined. It has been determined that for some substances, such as, for example, iron, cobalt and nickel, the angular momentum belonging to one Bohr magneton is

$$\frac{1}{2}\frac{h}{2\pi}, \tag{4.7}$$

where h is again the Planck constant.

This result is rather surprising; namely, if the magnetic moment is produced by the electrons revolving in the atoms, the ratio of the angular momentum and magnetic moment should be $2m_0c/e$ instead of m_0c/e as found here. This can easily be demonstrated: The absolute value of the magnetic moment produced by a given current in a closed circuit is equal to the product of the current and the encircled area divided by c, if electrostatic units are used:

$$|\mu| = \frac{if}{c}. \tag{4.8}$$

If the current is produced by electrons revolving around the nucleus, then in the case of one electron i is equal to $-e/\tau$, if τ is the time of one revolution, thus the absolute value of μ is

$$| \mu | = \frac{1}{c} \frac{e}{\tau} f = \frac{1}{c} e \frac{f}{\tau}, \tag{4.9}$$

where f/τ is the area described per unit time by the radius vector drawn from the nucleus. Generalizing this relation for the case of more electrons, and using vector notation,

$$\mu = - \frac{1}{c} \sum_i e \frac{1}{2} [\mathbf{r}_i, \mathbf{v}_i] = - \frac{e}{2c} \sum_i [\mathbf{r}_i, \mathbf{v}_i], \tag{4.10}$$

where $\mathbf{r}_i$ is the radius vector of the ith electron, and $\mathbf{v}_i$ is its velocity. (The expression also gives the direction of the angular momentum correctly.)

The corresponding angular momentum is

$$\mathbf{N} = \sum_i m_0 [\mathbf{r}_i, \mathbf{v}_i] = m_0 \sum_i [\mathbf{r}_i, \mathbf{v}_i], \tag{4.11}$$

where m_0 is the electron mass. Thus $\mu = (e/2m_0c)\mathbf{N}$, that is, the ratio of the absolute values of the angular momentum and magnetic moments is in electrostatic units

$$\frac{|\mathbf{N}|}{|\mu|} = \frac{2m_0c}{e}. \tag{4.12}$$

As mentioned above half of this ratio was found for iron, cobalt and nickel. This is interpreted in the following way: for these substances the angular momentum is produced not by the revolution of the electrons around the nucleus, but each electron has its own angular momentum and magnetic moment, and in iron, cobalt and nickel the moments observed are the sum of these. The angular momentum of an electron is $\frac{1}{2} h/2\pi$, its magnetic momentum is $eh/4\pi m_0c$, that is just one Bohr magneton. This angular momentum is called *spin*. Formerly it was interpreted that the electron is turning around its axis (spinning), but this explanation had its difficulties, because in this case the peripheral velocity of the electron would be higher than the velocity of light. In the following the electron spin and the magnetic moment associated with it will be regarded as the characterisic property of an electron, that is just such an important characteristic constant of the electron as its charge or mass. Dirac has given a more profound interpretation of the spin in quantum mechanics.

In substances where

$$\frac{|\mathbf{N}|}{|\mu|} = \frac{2m_0c}{e},$$

the moments are produced by the revolution of the electrons around the nucleus, the spins and magnetic moment of the electrons compensate each other and do not contribute to N or μ. There are some substances, for instance some salts and rare earth metals, where the value of $|N|/|\mu|$ lies between

$$\frac{m_0 c}{e} \quad \text{and} \quad \frac{2m_0 c}{e},$$

this can be interpreted by assuming that the moments are vectorially composed of the moment due to the revolving motion and the spin of the electrons.

The hypothesis that the electrons possess a spin and an intrinsic magnetic moment was proposed by Goudsmit and Uhlenbeck, and turned out to be very fruitful. Not only the magnetomechanical anomaly became interpretable, but it proved to be essential in the theory of spectra as well. The experiments fully agreed with the hypothesis of Goudsmit and Uhlenbeck.

5. The Elements of Spectra

Besides the afore-mentioned experimental results the investigation of spectra with their high number of empirical results gives a further insight into the structure of the atoms.

The best known spectrum is that of the Sun. It can be produced by letting the sunshine fall through a narrow slit into a darkened room and setting a triangular glass prism into the light beam. In this case the following observation can be made: the white light beam spreads into a coloured one, and placing a piece of white paper into the light path, a coloured band, a spectrum is observed in which the sequence of the colours is similar to that of the rainbow, i.e. at one end one finds the red, followed by the orange, yellow, green, blue and violet colours. Such a spectrum where the colours line up continuously is called a continuous spectrum. The colour sensation is evoked in this case by the different monochromatic parts of the spectrum, that is by the electromagnetic radiations of different wavelengths. Thus, for instance, in the spectrum of the Sun the wavelength of the electromagnetic radiation decreases from the red to the violet. The wavelength of the red light is 7.5×10^{-5} cm, that of the violet is 3.6×10^{-5} cm. The visible part of the Sun's spectrum is only a small part of the total spectrum which extends beyond the red into the infrared, beyond the violet into the ultraviolet spectrum range. The ultraviolet and infrared radiations are also of electromagnetic character, the only difference between them and the visible one is that for the former the wavelength is shorter, for the latter longer than for light. A very broad range of electromagnetic radiation is known; thus for instance X-rays with a wavelength of 10^{-6}–10^{-9} cm and γ-radiation with a wavelength of 10^{-9}–10^{-11} cm.

Incandescent solids and liquids emit spectra similar to the Sun's spectrum. The spectra of gases and vapours are quite different, they are not continuous, but are composed of single lines, that is in the spectrum emitted by vapours and gases only discrete wavelengths are present. The investigation of these line-spectra gave the most important information on the structure of the atoms. The investigation of the spectra and the determination of the wavelengths of the spectral lines can be carried out with the help of prism and grating spectroscopes. The nature and working principles of these instruments can, however, not be discussed here.

At the end of the last century Balmer noticed that the wavelengths of the four lines of the H-spectrum known in those days can be described by a simple formula. Later this was brought to the following very simple form:

$$\bar{v} = R \left(\frac{1}{2^2} - \frac{1}{n^2} \right), \tag{5.1}$$

where $\bar{v}$ is the number of waves per cm, that is the wave number. Thus if λ is the wavelength, $\bar{v} = 1/\lambda$. R is a constant, the value of which is, according to recent determinations

$$R = 109737.309 \text{ cm}^{-1}. \tag{5.2}$$

R is called the Rydberg constant. n can be an integer between 3 and 6. By substituting the integers into the equation the wave numbers of the four then known lines of H were attainable, these were the so-called H_α, H_β, H_γ and H_δ lines. It turned out later that by means of equation (5.1), the so-called Balmer formula, not only the four H-lines can be created but substituting the integers $n = 7, 8, 9, \ldots$, etc., the wave numbers of further lines of the H-spectrum can be obtained. These lines lie in the ultraviolet spectrum. At present more than thirty lines are known, the wave numbers of which are generated with a very high precision by the Balmer formula.

From the investigation of the line-spectra it became clear that the spectral lines of an element can be ordered in different groups, called series. It is characteristic for these series that in a given series going towards smaller wavelengths (higher wave numbers) the distance between the lines of a given series becomes smaller, and converges towards a well-defined boundary. The value of this boundary is characteristic for each series. The lines belonging to a given series can be recognized by their physical nature (the sharpness, simplicity or multiplet structure, their behaviour in a magnetic field, etc.).

16

Later further series were found in the H-spectrum, the lines of which could be produced by a formula very similar to (5.1):

$$\bar{v} = R \left(\frac{1}{s^2} - \frac{1}{n^2} \right).$$ (5.3)

If $s = 1$ and $n = 2, 3, 4, \ldots$ it is called the Lyman series,
if $s = 3$ and $n = 4, 5, 6, \ldots$ it is called the Paschen series,
if $s = 4$ and $n = 5, 6, 7, \ldots$ it is called Bracket series.

Already these simple formulae show that the wave numbers of the H spectral lines can be produced as the difference of two expressions (R/s^2 and R/n^2); these expressions are called in spectroscopy: *terms*. But not only the wave numbers of the H spectral lines, but those of every element both in the optical and in the X-ray spectrum can be produced as differences of terms. This gives a far clearer order than a mere sequence of spectral lines. The determination of the terms is one of the most important tasks of spectroscopy. It can be seen from eq. (5.3) that for H the term corresponding to the limit of the spectrum is R/s^2, approaching this value if n increases beyond any limit.

The spectra of the alkali metals consist of similar series to that of H. The wave numbers of the spectral lines corresponding to these can be created as the difference of the members of different series of terms. Namely, the terms can be ordered into series; it is usual to distinguish $S, P, D, F, G, H, \ldots$ series of terms. (Every letter denotes a series.) Each term of a series is distinguished by a number written before the letter. Thus, for instance, $1S, 2S, 3S, 3P, 6P, 4D, 7D$. The wave numbers of the more important series can be produced by means of this notation as follows:

The so-called principal series: $\bar{v} = 1S - mP$, $m = 2, 3, 4, \ldots$
The first (diffuse) secondary series: $\bar{v} = 2P - mD$, $m = 3, 4, 5, \ldots$
The second (sharp) secondary series: $\bar{v} = 2P - mS$, $m = 3, 4, 5, \ldots$
The Bergmann (fundamental) series: $\bar{v} = 3D - mF$, $m = 4, 5, 6, \ldots$

There are some other series too. The spectra of the singly ionized alkali earths, of the doubly ionized alkali-earth metals, etc., proceeding in the periodic system from the left towards the right are exactly similar to those of the alkali metals.

The spectra of the alkali-earth metals are more complicated than those of the alkali metals, but the different series are well distinguishable also in these spectra. Proceeding in the periodic system from left towards right the spectra of the elements beyond the alkali-earth metals become more and more complicated, and it is difficult to recognize the different series,

17

but still, in principle it is possible to produce the lines by means of terms, but in practice it is often difficult to determine the terms.

In several spectra the lines are not single, but consist of some components lying very close to each other. It is usual to distinguish between the lines according to their structure: the single lines are called singlets; there are lines composed of two parts, these are the doublets; there are triplets, etc. The multiplicity of the lines has to be due to the multiplet structure of the terms generating them. Accordingly the terms are called singlet, doublet, triplet, etc., terms. In the case of the alkali metals the S-terms are all simple, singlet ones, the other alkali terms are doublets.

In Chapter 2 we will discuss in detail how Bohr succeeded in interpreting on the basis of quantum mechanics the very rich experimental material gained by the investigation of the spectra, and how this interpretation led to further theoretical conclusions. Before it, however, we have to deal with the photoelectric effect and with the de Broglie waves. These phenomena prove the dual, corpuscular and undulatory character of light and matter.

6. The Photoeffect, the Photon

In the first section the Hallwachs or photoelectric effect has already been mentioned briefly. The basic experimental effect of this phenomenon is the following: If light, ultraviolet radiation, X-rays or γ-rays fall on a metal plate, electrons are ejected from this plate and the metal becomes positively charged. The velocity of the electrons can be determined. There exists a very important relation between the velocity v of electrons with the highest velocity, and the frequency v of the light falling on the plate:

$$hv = \frac{1}{2} m_0 v^2 + A , \qquad (6.1)$$

where h is the Planck constant, m_0 is the mass of the electron, A is a constant characteristic of the metal and independent of the frequency. A is the work the electron has to perform when stepping out of the metal, therefore it is called *work function*.

Equation (6.1) is the basic equation of the photoelectric effect first formulated by Einstein.

At high frequencies, in the range of X-rays and γ-rays, A can be neglected because v and v are so big that $A \ll \frac{1}{2} m_0 v^2$, and thus

$$hv = \frac{1}{2} m_0 v^2 . \qquad (6.2)$$

The velocity of the ejected electrons does not depend on the intensity of the radiation, only on v. It is only the number of emitted electrons that depends on the intensity.

The photoelectric effect can be observed for insulators as well, but in this case the measurement is much more complicated.

The theoretical interpretation of the photoelectric effect encountered difficulties and led to quite new concepts. The problem was the following: the electron attains its kinetic energy as well as the energy necessary for leaving the substance from the light wave. Thus in the case of low light levels — if the light energy is distributed uniformly over the light waves — it would take very long (for very low light levels even days) till enough energy for leaving accumulates on the area of an atom (i.e. a disc with a diameter of roughly 10^{-8} cm). In contradiction to this the experiments show that even for weak light intensities the photoeffect sets in instantly.

That is why Einstein made the basic assumption that the energy is not continuously distributed in the light waves (and generally in all other forms of electromagnetic radiation) but is concentrated into packets, so-called photons. If a photon bumps against an atom, it passes its energy to an electron of the atom and the photon ceases to exist.

According to (6.1) the energy E of the photon can be supposed to be

$$E = hv . \tag{6.3}$$

According to the theory of relativity the mass belonging to the energy E is

$$m = \frac{E}{c^2}, \tag{6.4}$$

thus in our case

$$m = \frac{hv}{c^2} . \tag{6.5}$$

From this the absolute value of the momentum of the photon is

$$p = mc = \frac{hv}{c} = \frac{h}{\lambda} . \tag{6.6}$$

From this the wavelength of the light can be expressed by the help of the momentum of the photon

$$\lambda = \frac{h}{p} . \tag{6.7}$$

Basic entities of the photons are their energy (6.3) and their momentum (6.6).

By the help of the concept of photons the interpretation of the photo-effect is straightforward, as also in the case of low light levels the photo-effect occurs spontaneously if it is supposed that the energy is concentrated in photons.

On the other hand the wave-nature of light cannot be dispensed with either, because from interference phenomena it is clear that along the light ray the electric and magnetic field changes periodically. Thus light has a double, or dual nature, it is at the same time wave and particle. Which of these is seen in a given instance depends on the experiment. The inter-relation between the photons and the light wave is as follows: the number of photons is proportional to the intensity of the wave, that is to the square of the wave amplitudes. In a diffraction experiment, for instance, more pho-tons reach such a place where the intensity is higher than another one where the intensity is small; regions with zero intensity are not reached by photons. The intensity of the wave determines the distribution of the photons.

The photon concept turned out to be very fruitful. Thus, for instance, it was possible, using this concept to explain the Doppler effect too, without wave theory. It became possible to explain the red-shift of light reaching us from stars with a big mass. Another phenomenon, the explana-tion of which was a great success of the photon theory, was the Compton effect. If an electromagnetic ray is scattered by a substance the wavelength of the scattered light changes, becoming longer. This change depends only on the angle of scattering, and is independent of the scattering material and the wavelength of the incident radiation.

The wavelength shift is

$$\Delta\lambda = 2\lambda_0 \sin^2 \frac{\vartheta}{2}, \qquad (6.8)$$

where ϑ is the angle of deviation, and λ_0 is a universal constant:

$$\lambda_0 = 2.4 \times 10^{-10} \text{ cm} = 0.024 \text{ Å} \quad (1\text{Å} = 10^{-8} \text{ cm}). \qquad (6.9)$$

It is very easy to explain this phenomenon on the basis of the photon theory if one supposes that the mechanism of photon-scattering consists of the single act of photon–electron collisions, with the photon deviated from its original direction. This theory leads to equations which are in perfect agreement with experiment.

At first sight the dual nature of light as introduced here seems to be very strange but according to our present-day opinion it is a fundamental law and, as will be seen later, is valid also for material particles.

7. Matter Waves

In 1927 Davisson and Germer discovered a phenomenon of basic importance: rays consisting of electrons (electron rays) show similar characteristics as electromagnetic waves: they can be diffracted. This can be demonstrated by the help of the following experiment:

An electron ray is incident on a metal lattice, in a manner similar to that used by Thibaud for X-rays, i.e. nearly tangentially. In this case a diffraction pattern is observed, similar to that observed in the case of X-rays. A glass grating is not practical because it charges and deviates the electrons. From the diffraction patterns the wavelength could be determined. The following relation was found:

$$\lambda = \frac{h}{m_0 v} = \frac{h}{p}, \tag{7.1}$$

where h is the Planck constant, m_0 is the mass, v the velocity and p the momentum of the electron. The expression is similar to that found in case of photons [see (6.7)]. λ is the de Broglie wavelength of the radiation.

Thus similarly to the dual nature of light which shows besides the wave nature a corpuscular one too, the electron rays show besides their corpuscular character a wave character as well. If one places a thin crystal plate into the electron beam so that the electrons can penetrate through the crystal, a diffraction pattern, similar to that found with X-rays, will be observed. That means one receives the analogy of the Laue diagrams. If the electron beam traverses a microcrystalline (or polycrystalline) metal-foil, very well-defined Debye–Scherrer fringes will appear.

It has still to be proved that the diffraction is really produced by electrons and not by an electromagnetic radiation following the electron beam. This can be proved by demonstrating that a magnetic field displaces the pattern.

In the case of high-velocity electrons (they were accelerated by a 30,000 V potential difference) an exact correspondence with (7.1) was received. In the case of slow electrons (accelerated by a few hundred volts) some deviation from (7.1) was found. This discrepancy was explained by supposing that the refractive index in the crystal is different from that in air, that is the wavelength is different inside and outside the crystal.

Stern proved that a He atom beam possesses a wave nature too. The relation between the de Broglie wavelength and the momentum of the He atoms was the same as found for electrons. The 7400 times higher mass of the He atoms is compensated by the much lower velocity, thus λ is again of such a magnitude that the diffraction takes place on crystal lattices.

According to these it can be taken for certain that the corpuscular rays show in general – that is, not only in the cases mentioned – a wave nature. This duality is one of the most fundamental ideas of modern physics, and it belongs to the most important experimental basis of quantum mechanics.

CHAPTER 2

Bohr's Atom Model

8. The Bohr Model of the Hydrogen Atom

Rutherford's atom model proposed by Rutherford for the interpretation of the scattering of α-particles, as well as the photon hypothesis of Einstein were — in bold outline — the basis for Bohr in building his theory. These theories have already been dealt with in the preceding sections.

Bohr started directly from Rutherford's model according to which the electrons of an atom revolve around the nucleus like planets. This introduces the first difficulty. If the electrons revolve around the nucleus, according to classical electrodynamics, the atoms should emit electromagnetic radiation, because the nucleus–revolving-electron system generally possesses a changing electric moment (this is specially obvious in case of a single electron). Due to the electromagnetic radiation of the atoms the electrons would lose energy and thus after a given time would, while moving along a spiral path, fall into the nucleus. This means that on the basis of the classical theory the Rutherford model is unstable. The model had to be modified profoundly, and this was done by Bohr. According to Bohr's supposition the electrons can revolve around the nucleus only in definite orbits. In contrast to electrodynamics, Bohr supposed that if the electrons revolve in such orbits the atom does not emit. In each of these orbits the electrons possess an absolutely fixed amount of energy, and the electron can take up only energies corresponding to these stationary orbits. No continuous transition exists between these values of energy just as there is no continuous transition between the different orbits. That means that the atom can change its energy only abruptly while one of its electrons moves from one possible orbit to another. From this it follows that the atom cannot take up, i.e. absorb, energy *ad libitum* but only by exactly determined quantities. Thus, for instance, the atom can absorb light only if the energy $h\nu$ of the photon is equal to the difference of the two energy states of the atom and the energy of the photon increases the energy of the atom in the form of lifting an electron of the atom which possesses the proper energy into a state of higher energy, lying above the level by $h\nu$. During the opposite process, as the electron of an atom moves from a higher state of energy into a lower one, the energy difference is emitted by the atom in the form of an $h\nu$ energy quantum of electromagnetic radiation.

This can be expressed in the following form:

$$E_n - E_s = h\nu .\tag{8.1}$$

From this one reaches the following expression for the frequency or wave number of absorbed or emitted radiation

$$\nu = \frac{E_n - E_s}{h}, \qquad \bar{\nu} = \frac{\nu}{c} = \frac{E_n - E_s}{hc} .\tag{8.2}$$

Expression (8.1) is of fundamental importance in this theory and is called *Bohr's frequency relation*. It has already been encountered in the discussion of the Franck–Hertz experiment.

Bohr worked out his model first for the simplest atom, for H, where only one electron revolves around the nucleus. In this case the possible orbits and energies are obtained very simply. The electron is bound to the nucleus by the electrostatic Coulomb attraction. If only circular orbits are taken into consideration, and r is the radius of the orbit, then the absolute value of this force is e^2/r^2, where e is the positive elemental charge. The centrifugal force on the electron has an opposite direction, its absolute value is $m_0 v^2/r$, where m_0 is the electron mass and v its velocity. In a stationary orbit these two forces have to be equal, thus

$$\frac{e^2}{r^2} = \frac{m_0 v^2}{r} .\tag{8.3}$$

In order to calculate the orbit and energy of the electron another equation is needed. This is furnished by *Bohr's quantum condition*, an assumption defining the possible orbits and energies for the case of H. This assumption which can neither be derived from nor deduced to some other principle states for cases of circular orbits that the angular momentum (being for circular orbits equal to $m_0 v r$) has to be equal to an integer multiple of $h/2\pi$, thus

$$m_0 v r = n \frac{h}{2\pi} ,\tag{8.4}$$

where n is an integer. This integer is called a *quantum number* ($n \neq 0$).

The quantum condition can be formulated in a more general form as well:

$$\oint p_i dq_i = n_i h ,$$
$$(i = 1, 2, \ldots, f),\tag{8.5}$$

where the q_i are the generalized position coordinates of the system, and the p_i are the corresponding conjugate momenta, the definition of which is

$$p_i = \frac{\partial E_k}{\partial \dot{q}_i}$$

where E_k is the kinetic energy. Generally the number of quantum conditions is equal to the number of generalized coordinates. In the case of circular orbits $E_k = \frac{1}{2} m_0 r^2 \dot{\phi}^2$, thus one obtains $p = m_0 r^2 \dot{\phi} = m_0 r v$ (because $r\dot{\phi} = v$). Thus in the case of circular orbits the general form of the quantum condition is as follows:

$$\oint p\,dq = m_0 v r \oint d\phi = 2\pi m_0 r v = nh, \tag{8.6}$$

which is the same as (8.4).

By means of (8.3) and (8.4) the radius of the different possible circular orbits and the velocities in them can be calculated. Squaring (8.4) and multiplying it by (8.3) one obtains

$$m_0^2 r^2 v^2 \frac{e^2}{r^2} = n^2 \frac{h^2}{4\pi^2} \frac{m_0 v^2}{r}; \tag{8.7}$$

from this the radii of the circular orbits are

$$r_n = \frac{h^2}{4\pi^2 m_0 e^2} n^2, \tag{8.8}$$

where the radii belonging to different values of n are distinguished from each other by the index n. By substituting this value of r_n into (8.4) one gets the electron velocity in the nth orbit:

$$v_n = \frac{2\pi e^2}{h} \frac{1}{n}. \tag{8.9}$$

Thus the radii of the orbits are proportional to the squares of the integers. The electron velocities corresponding to these orbits are inversely proportional to the integers. The smallest orbit, where $n = 1$, has

$$r_1 = \frac{h^2}{4\pi^2 m_0 e^2} = 0.529 \times 10^{-8} \text{ cm}. \tag{8.10}$$

In this orbit the velocity of the electron is $v_1 = 2\pi e^2/h = 2.2 \times 10^8$ cm s^{-1}.

25

The electron energy, composed of the potential and kinetic energy, is of great importance

$$E = - \frac{e^2}{r} + \frac{1}{2} m_0 v^2 . \tag{8.11}$$

From (8.3), if it is multiplied by $r/2$, onens obtains $\frac{1}{2} (e^2/r) = \frac{1}{2} m_0 v^2$. By means of this, one gets from (8.11):

$$E = - \frac{1}{2} \frac{e^2}{r} . \tag{8.12}$$

From (8.8) the values of r can be written into (8.12). In this case, by using the indexes for the energy values as well, one gets

$$E_n = - \frac{2\pi^2 m_0 e^4}{h^2} \frac{1}{n^2} . \tag{8.13}$$

This expression provides the possible energy values of the H-atom which correspond to electron orbits of different radii. On the basis of Bohr's frequency relation we learn from this formula that when the electron makes a transition from the orbit with the quantum number n to an orbit with the smaller quantum number s, the wave number of the emitted light is

$$\bar{\nu} = \frac{E_n - E_s}{hc} = \frac{2\pi^2 m_0 e^4}{h^3 c} \left(\frac{1}{s^2} - \frac{1}{n^2} \right) . \tag{8.14}$$

By inserting the values of the atomic constants, for the factor

$$\frac{2\pi^2 m_0 e^4}{h^3 c}$$

it was found that within the limits of a small difference (this will be dealt with shortly) it corresponds to the value of the experimentally found Rydberg constant. That is, formula (8.14) deduced by Bohr is equal to the experimental formula; that is, one gets for R the following formula:

$$R = \frac{2\pi^2 m_0 e^4}{h^3 c} , \tag{8.15}$$

which contains only atomic constants. Thus Bohr's theory dealing with the H-atom was able to explain the H-spectrum and the terms. The single series are obtained from (8.14) by putting one of the integers 1, 2, 3, 4 in the place of s, and n has to be an integer higher than s. This corresponds to the fact that, for instance, during the emission of light, the electrons of the atom show a transition from a higher energy level to a lower one. The lowest energy level corresponds to $n = 1$. In this case, as we have already seen, the radius of the orbit is the smallest. This is called the *ground state* of the H-atom. The energy of the electron, that is of the atom is in this case, as seen from (8.13): $E_1 = -Rhc = -13.54$ eV. The absolute value of this energy is called the ionization potential of the H-atom, because this amount of energy is required for ejecting the electron which is in the ground state of the atom, from the atomic core, that is to ionize the atom.

This theory can easily be adopted for the case when the charge of the atom core is in general $+Ze$ and a single electron revolves around it. This is the case, for instance, for singly ionized He, doubly ionized Li, etc. The calculation can be performed in a similar way, only in the formula of the energy and $\bar{v}$ the factor Z^2 has to be added, according to which the spectral lines shift towards the ultraviolet. Obviously the formulae for r_n and v_n change as well.

It has already been mentioned that a small difference exists between the theoretically deduced and experimentally determined values of R. The interpretation of this was a great success of Bohr's theory. Up to this point the nucleus was regarded as a stationary object; in the case of more accurate calculations, however, one has to take the movement of the nucleus into consideration: the nucleus is revolving around the stationary centre of mass. The calculations are simple and lead to results that R is replaced by the following factor:

$$\frac{R}{1 + (m_0/M)}, \tag{8.16}$$

where m_0 is the mass of the electron and M that of the nucleus. Thus, fo instance, the Rydberg constant of H is

$$R_{\mathrm{H}} = \frac{R}{1 + (m_0/M_{\mathrm{H}})}.$$

That for singly ionized He

$$R_{\mathrm{He}} = \frac{R}{1 + (m_0/M_{\mathrm{He}})}.$$

27

If the movement of the nucleus is taken into consideration, instead of R factor

$$\frac{R}{1 + (m_0/M)}$$

has to be written into the formulae giving the wave numbers of the spectral lines. The measurements fully confirm this.

As long as only circular orbits are considered the quantum state of H can be characterized by a single quantum number. If, however, elliptic orbits with the nucleus in one of the focal points are allowed too, then, as pointed out by Sommerfeld, two quantum conditions and consequently two quantum numbers occur because in this case two coordinates are necessary to define the position of the electron; ϕ the azimuth and r the distance of the electron from the nucleus are these coordinates. If the generalized momenta are p_ϕ and p_r, the quantum conditions can be written in the following form:

$$\oint p_\phi d\phi = kh, \tag{8.17}$$

$$\oint p_r dr = n_r h, \tag{8.18}$$

where k and n_r are integers, k is the so-called *azimuthal*, n_r the *radial quantum number*. The integration has to be extended over a full revolution of the electron. According to (8.17) the absolute value of the angular momentum has to be an integer multiple of $h/2\pi$, that is the absolute value of the angular momentum is $k(h/2\pi)$. The following condition holds for these quantum numbers: the radial quantum number n_r can be equal to zero or any integer, the azimuthal quantum number k, however, cannot be equal to zero. To stay in agreement with the experimental findings instead of k the orbital quantum number l had to be introduced. This is an integer, by 1 smaller than k, that is $k = l + 1$; thus l can be equal to zero as well. According to experimental facts the absolute value of the angular momentum is $l(h/2\pi)$ and not $k(h/2\pi)$.

$$n = n_r + k = n_r + l + 1 \tag{8.19}$$

is called the principal quantum number. For a given principal quantum number n, the highest value of k is n, and the highest value of l is $n - 1$.

In case of a circular orbit the quantum conditions (8.17) and (8.18) reduce to the equation (8.6); as in this case $p_r = 0$, and thus only (8.17) remains. That means that in case of a circular orbit $k = n$.

28

In the case of a nuclear charge Ze one gets for the semi-major and semi-minor axes of the elliptic electron orbit

$$a = \frac{n^2}{Z} a_H, \qquad b = \frac{nk}{Z} a_H, \tag{8.20}$$

where a_H is the radius of the innermost (circular) H-orbit. Also from these equations it is evident that when $k = n$ the electron orbit is circular. If $k \neq n$, the orbit is elliptic, which becomes more and more elongated as k becomes smaller and smaller compared to n. Figure 5 shows the orbits belonging to $n = 3, k = 1; n = 3, k = 2$ and $n = 3, k = 3$. The length of the major axes of the ellipses belonging to a given principal quantum number n are equal to each other and are equal to the diameter of the corresponding circular orbit.

It can be proved that for the H-atom, as long as no relativistic corrections are introduced, in the case of elliptic orbits the energy depends only on $n = n_r + k$, that means one receives the same energy levels as in the case of circular orbits.

Up to now the fine structure of the H-spectrum has been neglected. This phenomenon is the following: some lines of the H-spectrum turn out to show a fine structure if investigated under very high resolution. This was interpreted by Sommerfeld applying relativistic mechanics in his calculations. He concluded that the electron orbits revolve around their focal point, that is the electron moves along an orbit as shown in Fig. 6. For the energy he received the important result that it depends not only on $n = n_r + l + 1$, but also separately on l, that means that the energy levels which coincided in the preceding calculation break up into a multiplicity of levels resulting in the fine structure of the H-spectrum. Sommerfeld's theory is, however, incomplete. The correct treatment of the fine structure of the H-spectrum became possible only through Dirac's quantum theory.

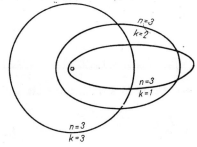

FIG. 5. Bohr electron orbits in H-atom for the principal quantum number $n = 3$

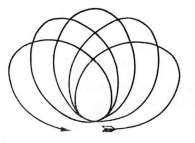

FIG. 6. One of the electron orbits of the H-atom according to the relativistic Bohr–Sommerfeld theory

9. Moseley's Law

An atom with an atomic number Z, if it lost all but one of its electrons, would emit light, the wave number of which could be described — if one neglects the movement of the nucleus — by the following formula:

$$\bar{v} = RZ^2 \left(\frac{1}{s^2} - \frac{1}{n^2} \right). \tag{9.1}$$

Due to the Z^2 factor the lines for $Z = 30$ fall already into the X-ray part of the spectrum. Structures ionized 30 times or even to a higher extent cannot be produced experimentally. In the X-ray spectrum, however, one finds lines which can be generated by a formula that is analogous to the above-mentioned one and can be deduced from it.

Before deducing this formula first it has to be mentioned that not all electrons of an atom with a higher atomic number are in the lowest possible state of energy, i.e. in the quantum state $n = 1$. The electrons are distributed in different quantum states. Those with equal principal quantum states form groups. They are arranged into so-called shells, which are identified by the letters $K, L, M, N, \ldots$ according to their increasing distance from the nucleus, and incorporate the electrons being in the states corresponding to the principal quantum numbers $n = 1, 2, 3, 4, \ldots$

If an atom loses an electron from its inner shell due to a high-energy photon or to a collision with an electron, an electron of higher energy from another shell can take its place, while the atom will emit, in the case of high atomic numbers, X-rays. If the interaction of electrons is neglected, the wave number of the emitted X-rays could be determined by the above-mentioned formula. In reality, however, the inner electrons of the atom decrease the electrostatic influence of the nucleus, and so for every shell a constant Z_n can be introduced which accounts for this decrease: in a given shell the electrostatic potential of the nucleus is not

$$\frac{Ze}{r} \quad \text{but} \quad \frac{(Z - Z_n)e}{r}.$$

According to this the energy of an electron being in a quantum state with principal quantum number n is

$$E = (Z - Z_n)^2 \frac{Rhc}{n^2}. \tag{9.2}$$

This reasoning is of interest mainly because Z_n proved to be largely independent of Z. For the K and L-shells Z_n can be taken equal to unity, and thus the wave number of the line for the transition from the L-shell to the K-shell is

$$\bar{\nu} = (Z - 1)^2 R \left(\frac{1}{1^2} - \frac{1}{2^2} \right) = \frac{3}{4} (Z - 1)^2 R . \tag{9.3}$$

This is the famous Moseley law, according to which the wave number of the K_α-line is proportional to $(Z - 1)^2$. The elements between $Z = 4$ and $Z = 92$ obey this law very accurately. The agreement is highest for elements with medium Z values, as Z_n varies slightly with Z. By the help of the Moseley law the atomic number of an element can be given immediately, if the wave number of the K_α-line is known. The X-ray spectra will not be dealt with here; we only mention that their theory is complicated.

10. Atomic Spectra of the Alkali and Similar Atoms

Starting from the quantum conditions, it was possible to extend Bohr's theory of the H-atom to other more complicated atoms, in the first place to the alkali atoms, where one of the outer electrons is much more weakly bound than the others. This electron, which is usually farther from the nucleus than the other ones, interacts with the joint field of the nucleus and the other electrons. Sufficiently far from the nucleus this effect of the nucleus and of the inner electrons is very similar to the effect of a point charge, thus to the effect of a H-nucleus, and so the terms corresponding to these states agree nearly totally with the H-terms. But if the outermost electron (the so-called *valence electron*) gets nearer to the nucleus, or even penetrates in between the inner electrons, the joint field of the nucleus and the inner electrons differs from the field of a point charge, and so the terms corresponding to these states differ from the H-terms to a greater extent. Rydberg introduced the terms of the alkali atoms, to a first approximation, in the following form:

$$T = \frac{R}{(n + \delta_1)^2} ; \tag{10.1}$$

Ritz gave, as a second approximation, the following form:

$$T = \frac{R}{\left(n + \delta_1 + \dfrac{\delta_2}{n^2} \right)^2} , \tag{10.2}$$

where δ_1 and δ_2 are the so-called Rydberg, and Ritz corrections, respectively; which depend, in their turn, only on k and l. The terms thus differ from the H-terms, because in the denominator the square of $n + \delta_1$ and $n + \delta_1$

31

$+ (\delta_2/n^2)$ stands instead of the square of n, and these are no longer integers. The terms of the alkalis are often written also in the following form: R/n^{*2} where n^{*2} stands instead of the denominator of the former expression. n^* is called the *effective quantum number*; it is, however, no longer an integer. δ_1 and δ_2 can be determined on the basis of the Bohr theory and the results are generally in good agreement with the observations. There are two reasons why the alkali terms do not agree with the H-term. Firstly, as already mentioned, in states when the valence electron gets near to the nucleus the field is not Coulombian, and secondly the valence electron, even if far from the nucleus, polarizes the nucleus–core-electron system (the alkali ion). Due to this the valence electron extends the system of the positive nucleus and the negative electrons because it attracts the nucleus and repels the electrons, and so produces an electric dipole, the effect of which has also to be taken into consideration.

The alkali terms, as seen above, depend not only on $n = n_r + l + 1$, but depend separately on l as well (because δ_1 and δ_2 depend on l). The terms corresponding to a given l value are grouped into series, the members of which differ in n. The terms corresponding to $l = 0, 1, 2, 3, \ldots$ are called $S, P, D, F, \ldots$ The principal quantum number, n, is written before the sign of the term. Thus, for example, the members of the S-term series are the following: $1S, 2S, 3S, \ldots$, as was briefly mentioned in Section 5.

By going in the periodic chart from left to right, terms similar to the alkali ones are found for the singly ionized alkali earths, the doubly ionized alkali-earth metals, etc. The structure of these ions is similar to that of the alkali atoms, only the influence of the nucleus and of the core electrons on the loosely bound electron can be described for these as the field of a doubly, triply, etc., charged ion, and so the terms have to be multiplied by $2^2, 3^2, \ldots$ for the different cases. Also these terms can be explained by Bohr's theory. In fact the theory can be expanded even to the case of more complicated atoms.

11. Correspondence Principle

Coming back to the H-atom, the frequency corresponding to a transition from a state with a principal quantum number m to one with the principal quantum number n if $m - n \ll n$ can be determined. In this case

$$v = Rc \left(\frac{1}{n^2} - \frac{1}{m^2} \right) = Rc \frac{m^2 - n^2}{m^2 n^2} = Rc \frac{(m + n)(m - n)}{m^2 n^2} \approx Rc \frac{2}{n^3}(m - n),$$

(11.1)

as according to the condition $m \approx n$.

The classical frequencies in an orbit with the principal quantum number n are

$$\nu = \frac{v_n}{2\pi r_n} = \frac{(2\pi e^2/h)(1/n)}{(h^2/2\pi m_0 e^2)n^2} = \frac{2\pi^2 m_0 e^4}{h^3} \frac{2}{n^3} = Rc\frac{2}{n^3}. \tag{11.2}$$

Comparing this expression with (11.1) it can be seen that in case of large quantum numbers the frequency corresponding to the $m - n = 1$ quantum change leads to the twofold classical frequency, etc. That means that an asymptotic relationship holds for the quantum-theoretical and the classical frequencies. This is a special case of a general regularity, Bohr's so-called *correspondence principle*. According to this theory in limiting cases the quantum-theoretical laws generally pass over into their classical mechanical equivalents. The correspondence principle proved to be very important in quantum theory, as, for instance, by means of this principle it became possible to determine the intensity of the different lines corresponding to given quantum changes and to determine the polarization of the radiation.

By means of the correspondence principle it was also possible to explain why some changes of quantum numbers do not occur, and hence why the lines corresponding to these transitions cannot be seen in the spectrum. Thus the so-called selection rules could be defined. These determine the allowed transitions between the quantum states, i.e. which terms combine with each other. Such selection rules are the following: only such terms combine with each other where the change of l is one ($\Delta l = \pm 1$), no other combination is possible. Thus, for instance, an S-term can combine with a P-term, but cannot combine with another S-term or with a D-term. There is no selection rule for the principal quantum number n, and so transitions are allowed between states belonging to whichever principal quantum number if the transition between the states is otherwise allowed. Two other selection rules will be discussed later (see Section 29).

12. The Stark and Zeeman Effects

Here we are concerned with the Stark and Zeeman effects only briefly. If an atom is placed in an electric field its spectral lines usually are split into several components or are displaced. This is the Stark effect. The theory accounts for this, namely the different terms undergo a change in a strong electric field and the exactly coinciding terms split into several different terms, leading to a splitting of the lines. The displacement of some lines is due to a change in the corresponding terms. The external field affects the intensity of the lines too.

Namely, as follows from the correspondence principle due to an external field some *forbidden transitions* (which do not fulfil the selection rules) become possible and this accounts for the occurrence of forbidden lines in the spectrum of emitting atoms in strong electric fields.

The basis of the Zeeman effect, which is so important in the understanding of the structure of the atoms, is the following. If the emitting atoms are placed in a magnetic field, their spectral lines split into several lines. The explanation of this effect is very similar to that of the Stark effect, that is; due to the magnetic field the terms normally coinciding split into several components and this leads to a splitting of the lines. Later this effect will be dealt with in detail, and here only the following result, being of great importance for the structure of the atoms, will be mentioned.

It appears from the behaviour of atoms in a magnetic field that the component of the orbital angular momentum, that is, the angular momentum, due to the revolution of the electron in the atom along the direction of the magnetic field, can take only well-defined values, integer multiples m of $h/2\pi$. m is called the *magnetic quantum number*. Thus, for instance, in the state corresponding to the $l = 4$ orbital quantum number the angular momentum of the electron can point, in respect to the external magnetic field, only into the directions shown in Fig. 7, that is the components of l in the direction of the magnetic field can only be 4, 3, 2, 1, 0, -1, -2, -3, -4. In this case the component can take on altogether only $2\times4 + 1 = 9$ values. Generally to a given orbital quantum number l the following magnetic quantum numbers belong: $l, l - 1, \ldots 1, 0, -1, \ldots, -(l - 1), -l$, that is $2l + 1$ values belong to l. That shows that the magnetic quantum number can be negative as well. As seen in Section 4, to a given orbital angular momentum N of the electron (i.e. angular momentum due to the revolution) a single magnetic moment

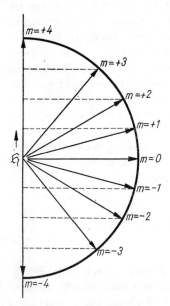

FIG. 7. The possible directions of the angular momentum for $l = 4$

$$\left(-\frac{e}{2m_0c}\,N\right)$$

belongs. The magnetic quantum number thus

describes how many times the component in the direction of the magnetic field of this magnetic moment is larger than

$$\frac{e}{2m_0 c} \frac{h}{2\pi},$$

which is 1 Bohr magneton.

From the correspondence principle one may deduce for m a selection rule, according to which transition are allowed only between such states where the change of m is

$$\Delta m = 0, \pm 1. \tag{12.1}$$

As the orbital angular momentum of the electron is perpendicular to the plane of the orbit, this result can be formulated in the following way: the plane of the electron orbit can be only in definite directions relative to the direction of the magnetic field.

13. Spin Quantum Number, Inner Quantum Number

For the full description of the different quantum states a further quantum number has to be introduced. As already mentioned, Goudsmit and Uhlenbeck introduced the very fruitful hypothesis of an intrinsic electron angular momentum. The value of this is, however, not an integer multiple of $h/2\pi$ but $\pm \frac{1}{2}$ times $h/2\pi$, i.e. $\pm \frac{1}{2}(h/2\pi)$. This quantum number of $\pm \frac{1}{2}$ is called *spin quantum number*, it is usually written as m_s. We know that a magnetic momentum $eh/4\pi m_0 c$ corresponds to the spin.

That means that a quantum state can be fully described by the quantum numbers n, l, m and m_s.

By using the spin the multiplicity of the terms can be explained. The spin, and in the case of more electrons the resultant of them can take up relative to the orbital angular momentum of the atom only a few well-defined directions. Due to an interaction between the magnetic moment corresponding to the spin moment and the orbital angular momentum the energy values are different for different orientations. This leads to the decomposition of the terms into several, lying near to each other in value, and this causes the multiplet structure of the lines.

Besides the quantum numbers mentioned, literature speaks about the so-called *inner quantum number* denoted by j. j gives the *total angular momentum* of the atom, which is the resultant of the orbital and spin angular momentum. j is an integer or a half-odd integer. A quantum state is fully described by the above-mentioned four quantum numbers, thus j has not to be taken into account. We only mention that the quantum states can be described by other quantum numbers as well, namely by n, l, j and m_j; where m_j is another magnetic quantum number.

In respect to the external magnetic field the total angular momentum j can take up only such directions that its component in the direction of the field is m_j times $h/2\pi$, where m_j can be integer or half-odd integer and can be zero only for integer j's. The following selection rules hold for j: $\Delta j =$ $= 0$, ± 1, except the $0 \to 0$ transitions.

14. The Hyperfine Structure

As mentioned above, the multiplicity of the terms and of the spectral lines could be interpreted by means of the electron spin.

Besides the normal multiplet structure the spectral lines show a further structure, the so-called *hyperfine structure*. This is also a form of splitting of the lines. The splitting is, however, much smaller than the usual multiplet splitting and can be observed only using spectroscopes of very high resolution.

The hyperfine structure has two reasons. One is an isotope effect. It is well known that a large number of elements are a mixture of isotopes with different masses. As already seen in the case of H, for an exact treatment it has to be taken into consideration that the electron revolves not around the stationary nucleus but the electron and the nucleus revolve around their joint centre of mass. Due to this the mass of the nucleus appears in the wave-number formula of the spectral lines. In case of isotopes this condition results in the splitting of the spectral lines. The frequency of the lines is influenced by the different structures of different isotopic nuclei as well.

Another reason for the hyperfine structure is that the nuclei also possess a magnetic moment and spin. The order of magnitude of the magnetic moment of the nucleus is equal to one nuclear magneton. This is roughly 1800 times smaller than the Bohr magneton. Due to the nuclear spin, similar to the electron spin, the spectral lines split once more but as the magnetic moment corresponding to the nuclear spin is much smaller than the Bohr magneton, the corresponding splitting is also much smaller.

15. The Energy Terms of Many-electron Configurations and their Symbols

In the following we are concerned briefly with the problem of how to combine the angular momenta and the corresponding magnetic moments in the case of many electrons. The possible quantum states and terms will be determined. For this reason a two-electron system will be discussed, where the existing regularities can be understood. Instead of dealing with the problem of putting together the angular momenta, the corresponding quantum numbers will be combined. This can be done because the quantum numbers charac- terize the angular momentum and the magnetic moment unambiguously. Let us take two electrons; let for one of them the orbital quantum number be l_1, the spin quantum number $m_{s1} = \pm \frac{1}{2}$; let for the other one the corre- sponding quantum numbers be l_2 and $m_{s2} = \pm \frac{1}{2}$. The angular momentum and the quantum numbers, respectively can be combined in different ways; in the following only the so-called normal or Russel–Saunders coupling will be discussed. According to this l_1 and l_2 have to be put together into a resultant L, m_{s1} and m_{s2} into a resultant S, the final resultant is obtained by combining L and S. The orbital quantum numbers have to be put together as vectors, their resultant can, however, be only an integer. Thus the maxi- mum of L is $l_1 + l_2$, its minimum is (if $l_1 > l_2$) $l_1 - l_2$. Negative values cannot occur, here they would be senseless, as there is no reference to what could be a positive or a negative direction. Thus, if for instance $l_1 = 3$ and $l_2 = 2$ then $L = 5, 4, 3, 2, 1$. The spins can be parallel or antiparallel to each other. In the antiparallel case $S = 0$, in the parallel one $S = 1$. In this latter case S can be oriented in three different ways compared to L: parallel, perpendic- ular or antiparallel. The final resultant gives J. If, for instance, $L = 3$, then for $S = 1$, $J = 4, 3, 2$. If $S = 1$, a triplet term system is obtained; if $S = 0$, as there is no spin splitting, the term system is singlet.

According to $L = 0, 1, 2, \ldots$ the terms are designated by $S, P, D, \ldots$. The principal quantum number is written before the symbol, and the mul- tiplicity of the term is written also before the symbol, as an upper index. To distinguish between the multiplet terms the inner quantum number (J) is written as a lower index behind the symbol. Thus, for instance, in the preceding example we had the following terms: $^3F_4, ^3F_3, ^3F_2$ and $^1F_3$; if the principal quantum number (which was omitted in the preceding exam- ple) is also added to the symbols, a term symbol obtains, for instance, the following form: $3^2P_{\frac{1}{2}}$. It has to be mentioned that the S-terms are always singlet ones, because the electron spin or their resultant cannot be oriented with respect to anything.

16. The Pauli Principle. The Structure of the Periodic System on the Basis of the Pauli Principle

The quantum numbers n, l, m and m_s determine completely the quantum state of the electron. The quantum states corresponding to the different values of l ($l = 0, 1, 2, 3, 4, \ldots$) and the electrons in these states are usually called by the letters s, p, d, f, g .... Thus, for instance, one speaks about $4s$ quantum states and electrons, respectively. The appropriate principal quantum number, n, is written before the symbols. In the above-mentioned example $n = 4$, $l = 0$, or for a $2p$ electron $n = 2$, $l = 1$.

It can easily be determined how many s, p, d, f, ... states belong to a given principal quantum number. We have seen that to every l $2l + 1$ values of m belong and to each of these a state belongs. Every state, however, doubles due to the two possible values of m_s. Thus the total number of the states belonging to an l value is $2(2l + 1)$. For a given principal quantum number the number of the s, p, d, f, ... states is 2, 6, 10, 14, 18, ... One obtains all the possible states belonging to a given principal quantum number, n, that is the states corresponding to all the possible l values, if one sums the $2(2l + 1)$ values for every l. All the possible l values for a given n are: 0, 1, 2, 3, ... $n - 1$. As

$$\sum_{l=0}^{n-1} 2(2l + 1) = 2n^2, \tag{16.1}$$

one obtains for a given principal quantum number n the number of the quantum states belonging to it to be $2n^2$. Thus for $n = 1, 2, 3, 4 \ldots$ the total number of quantum states is 2, 8, 18, 32, 50, ...

From the point of view of the atomic structure it is very important to know how the electrons distribute over certain quantum states. Pauli's principle gives the answer to this question; according to this only a single electron can be in a given quantum state, and it can never happen that all the four quantum numbers of two or more electrons are equal. This statement proved to be of outstanding importance in physics. Using this principle the periodic system can be built up. An atom in its ground state can be regarded as a system that consists of a nucleus and of electrons. These fill first the quantum states of lowest energy and then successively those with higher energy. The simplest atom is the H-atom, consisting of a singly charged H-nucleus (a proton) and a single electron. The lowest energy state of this electron is the $n = 1$, $l = 0$, $m_s = \pm \frac{1}{2}$, but it is uncertain which m_s state is realized. In its ground state the electron of the H is in a $1s$ orbit. The next element is He, consisting of a doubly charged nucleus and two electrons. These are in the $n = 1$, $l = 0$, $m = 0$, $m_s = +\frac{1}{2}$ and $n = 1$, $l = 0$, $m = 0$, $m_s = -\frac{1}{2}$ quantum states, which means the two electrons are in the two $1s$ states belonging to the $n = 1$ principal quantum number and so the states with $n = 1$ are filled. For the next eight elements, starting with Li and ending with Ne, the quantum states belonging to $n = 2$ fill in, first the $2s$ states and then the $2p$ ones, so that, for instance, Ne has two $1s$, two $2s$ and six $2p$ electrons. For the next eight elements (from Ne till Ar) the $3s$ and $3p$ states fill in, successively. In the next period there are eighteen elements, from K to Kr, and the ten $3d$, two $4s$ and six $4p$ states fill in. It is interesting to mention that the outermost electron of K is not in a $3d$ state, as could be expected from the building of the preceding periods, but in a $4s$ state. The filling in of the $3d$ states starts at Sc and lasts till Cu, whereas the filling in of the $4p$ states starts at Ga and finishes at Kr. The phenomenon observed here, namely that the d-states (states with higher orbital quantum numbers) fill in only later, repeats itself during the building up of the further periods. In the next period of again eighteen elements (between Rb and Xe) the $5s$, $5p$ and $4d$ states fill in. The next period contains thirty-two elements (from Cs till Em), and the $6s$, $6p$, $4f$ and $5d$ states fill in. The fourteen $4f$ states fill in only later, for the fourteen rare earths, where the $4f$ states fill in successively. For the last six elements of the periodic system the $7s$ and $6d$ states become partially filled in. The filling numbers of the quantum states for different atoms is seen in the Stoner table (see pp. 40–41).

Thus the periodic system can be built up. The interpretation of the number of elements in the single periods (2, 8, 18, 32, . . .) is also self-explanatory. They give the number of all possible states belonging to a certain given principal quantum number n, which is, as we saw, $2n^2$. Writing $n = 1, 2, 3,$ $4, \ldots$ the number of elements in the single periods is gained.

17. The Interpretation of the Chemical Nature of the Elements on the Basis of the Atomic Structure. Some Remarks about Chemical Binding

The chemical nature of an element is governed by the outer electrons of the atom, as in most of the chemical phenomena, e.g. in the chemical binding, practically only these electrons are important. This explains, for instance, the very similar characteristics of the rare-earth metals, in this case namely, the arrangement of the outer electrons is the same. It has already been mentioned that for these elements the inner $4f$ states fill up.

For the noble gases, staying at the end of the periods, it is characteristic that they possess eight outer electrons: two s- and six p-states (except He,

Period	Atomic number	Element	1s	2s 2p	3s 3p 3d	4s 4p 4d 4f	5s 5p 5d	6s 6p 6d	7s
I	1	H	1						
	2	He	2						
	3	Li	2	1					
	4	Be	2	2					
	5	B	2	2 1					
II	6	C	2	2 2					
	7	N	2	2 3					
	8	O	2	2 4					
	9	F	2	2 5					
	10	Ne	2	2 6					
	11	Na	2	2 6	1				
	12	Mg	2	2 6	2				
	13	Al	2	2 6	2 1				
III	14	Si	2	2 6	2 2				
	15	P	2	2 6	2 3				
	16	S	2	2 6	2 4				
	17	Cl	2	2 6	2 5				
	18	Ar	2	2 6	2 6				
	19	K	2	2 6	2 6	1			
	20	Ca	2	2 6	2 6	2			
	21	Sc	2	2 6	2 6 1	2			
	22	Ti	2	2 6	2 6 2	2			
	23	V	2	2 6	2 6 3	2			
	24	Cr	2	2 6	2 6 5	1			
	25	Mn	2	2 6	2 6 5	2			
	26	Fe	2	2 6	2 6 6	2			
IV	27	Co	2	2 6	2 6 7	2			
	28	Ni	2	2 6	2 6 8	2			
	29	Cu	2	2 6	2 6 10	1			
	30	Zn	2	2 6	2 6 10	2			
	31	Ga	2	2 6	2 6 10	2 1			
	32	Ge	2	2 6	2 6 10	2 2			
	33	As	2	2 6	2 6 10	2 3			
	34	Se	2	2 6	2 6 10	2 4			
	35	Br	2	2 6	2 6 10	2 5			
	36	Kr	2	2 6	2 6 10	2 6			
	37	Rb	2	2 6	2 6 10	2 6	1		
	38	Sr	2	2 6	2 6 10	2 6	2		
	39	Y	2	2 6	2 6 10	2 6 1	2		
	40	Zr	2	2 6	2 6 10	2 6 2	2		
	41	Nb	2	2 6	2 6 10	2 6 4	1		
	42	Mo	2	2 6	2 6 10	2 6 5	1		
	43	Tc	2	2 6	2 6 10	2 6 6	1		
	44	Ru	2	2 6	2 6 10	2 6 7	1		

the Chemical Elements (Stoner table)

Period	Atomic number	Element	1s	2s 2p	3s 3p 3d	4s 4p 4d 4f	5s 5p 5d	6s 6p 6d	7s
V	45	Rh	2	2 6	2 6 10	2 6 8	1		
	46	Pd	2	2 6	2 6 10	2 6 10			
	47	Ag	2	2 6	2 6 10	2 6 10	1		
	48	Cd	2	2 6	2 6 10	2 6 10	2		
	49	In	2	2 6	2 6 10	2 6 10	2 1		
	50	Sn	2	2 6	2 6 10	2 6 10	2 2		
	51	Sb	2	2 6	2 6 10	2 6 10	2 3		
	52	Te	2	2 6	2 6 10	2 6 10	2 4		
	53	I	2	2 6	2 6 10	2 6 10	2 5		
	54	Xe	2	2 6	2 6 10	2 6 10	2 6		
	55	Cs	2	2 6	2 6 10	2 6 10	2 6	1	
	56	Ba	2	2 6	2 6 10	2 5 10	2 6	2	
	57	La	2	2 6	2 6 10	2 6 10	2 6 1	2	
	58	Ce	2	2 6	2 6 10	2 6 10 1	2 6 1	2	
	59	Pr	2	2 6	2 6 10	2 6 10 2	2 6 1	2	
	60	Nd	2	2 6	2 6 10	2 6 10 3	2 6 1	2	
	61	Pm	2	2 6	2 6 10	2 6 10 4	2 6 1	2	
	62	Sm	2	2 6	2 6 10	2 6 10 5	2 6 1	2	
	63	Eu	2	2 6	2 6 10	2 6 10 6	2 6 1	2	
	64	Gd	2	2 6	2 6 10	2 6 10 7	2 6 1	2	
	65	Tb	2	2 6	2 6 10	2 6 10 8	2 6 1	2	
	66	Dy	2	2 6	2 6 10	2 6 10 9	2 6 1	2	
	67	Ho	2	2 6	2 6 10	2 6 10 10	2 6 1	2	
	68	Er	2	2 6	2 6 10	2 6 10 11	2 6 1	2	
	69	Tm	2	2 6	2 6 10	2 6 10 12	2 6 1	2	
VI	70	Yb	2	2 6	2 6 10	2 6 10 13	2 6 1	2	
	71	Lu	2	2 6	2 6 10	2 6 10 14	2 6 1	2	
	72	Hf	2	2 6	2 6 10	2 6 10 14	2 6 2	2	
	73	Ta	2	2 6	2 6 10	2 6 10 14	2 6 3	2	
	74	W	2	2 6	2 6 10	2 6 10 14	2 6 4	2	
	75	Re	2	2 6	2 6 10	2 6 10 14	2 6 5	2	
	76	Os	2	2 6	2 6 10	2 6 10 14	2 6 6	2	
	77	Ir	2	2 6	2 6 10	2 6 10 14	2 6 7	2	
	78	Pt	2	2 6	2 6 10	2 6 10 14	2 6 9	1	
	79	Au	2	2 6	2 6 10	2 6 10 14	2 6 10	1	
	80	Hg	2	2 6	2 6 10	2 6 10 14	2 6 10	2	
	81	Tl	2	2 6	2 6 10	2 6 10 14	2 6 10	2 1	
	82	Pb	2	2 6	2 6 10	2 6 10 14	2 6 10	2 2	
	83	Bi	2	2 6	2 6 10	2 6 10 14	2 6 10	2 3	
	84	Po	2	2 6	2 6 10	2 6 10 14	2 6 10	2 4	
	85	At	2	2 6	2 6 10	2 6 10 14	2 6 10	2 5	
	86	Rn	2	2 6	2 6 10	2 6 10 14	2 6 10	2 6	
	87	Fr	2	2 6	2 6 10	2 6 10 14	2 6 10	2 6	1
	88	Ra	2	2 6	2 6 10	2 6 10 14	2 6 10	2 6	2
VII	89	Ac	2	2 6	2 6 10	2 6 10 14	2 6 10	2 6 1	2
	90	Th	2	2 6	2 6 10	2 6 10 14	2 6 10	2 6 2	2
	91	Pa	2	2 6	2 6 10	2 6 10 14	2 6 10	2 6 3	2
	92	U	2	2 6	2 6 10	2 6 10 14	2 6 10	2 6 4	2

where there are no *p*-states, as $n = 1$), belonging to the same principal quantum number, are filled, that is the eight outer electrons fill all the possible *s*- and *p*-states, thus the noble gases have a so-called closed (*s*, *p*) electron shell. This electron configuration is very stable and explains the chemical inactivity of the noble gases. In the periodic chart the column of noble gases is followed by the monovalent alkali metals, they possess in addition to the stable noble gas-like electron configuration a further electron in an *s*-state. The next column consists of the bivalent alkali earth metals, where in addition to the stable noble gas-like electron configuration two further electrons revolve, etc. The halogens stand in the last column, their maximum valency is seven, they have seven outer electrons, that means in the shell one electron is lacking to form the next noble gas-like configuration. As we shall see the valency depends on the number of the outer electrons; this can be interpreted using the so-called spinvalence theory.

Just these outer electrons of the elements play an important role in chemical binding. One type of chemical binding, the so-called *heteropolar* or *ionic binding*, can be understood on the basis of the above-mentioned facts. We saw that the electron configuration of the noble gases is very stable, thus it is obvious that this configuration tries to build up in nature. This is performed by the elements standing in one of the first columns of the periodic chart by passing down their plus electron and by the elements standing in one of the columns just before the noble gases by taking up an electron. Thus, for example, the NaCl molecule consists of a singly charged positive Na ion and a singly charged negative Cl ion. This is due to the fact that the Na passes down its outermost loosely bound electron easily, whereby its electron configuration becomes similar to that of Ne; Cl, on the other hand, takes an electron up easily, changing its electron configuration to be similar to that of the Ar-atom. The binding forces of the molecule are explained by the attraction between the oppositely charged ions. The binding of the ionic crystals (for instance, NaCl or KCl) can be explained in the same way, the essential character of this binding is the same as in case of the molecule, and can be looked upon as a giant molecule. The attractive forces of a MgO crystal are interpreted similarly, this crystal consists of doubly charged positive Mg ions and doubly charged negative O ions. In this case the two outermost electrons of the Mg are transferred to the O.

There is also another group of materials where the binding cannot be explained by the above-mentioned mechanism. Such are, for example, the molecules of the H, O, or N gases: the H_2, O_2 and N_2 molecules, or, for instance, the diamond crystal that consists of C atoms. Here the building blocks of the molecules or crystals are neutral. This sort of binding is called *homopolar* or *valence binding*. The Bohr model could not explain this type of binding. The homopolar binding can be explained only by wave mechanics.

For the sake of completeness two further types of chemical binding have to be mentioned here: the *metallic* and the *van der Waals* or polarization *binding*. The metallic binding describes, as seen from its name, the attractive forces acting between the atoms of metals, thus, for example, the alkali metals and Cu, Ag, Au or Fe shows this type of binding. The van der Waals binding is a relatively weak binding produced by certain polarization forces, occurring also between neutral atoms or molecules. The solid noble gases belong, for example, to this class. Neither the metallic nor the van der Waals binding could be interpreted by Bohr's theory. Both of them were explainable only on the basis of wave mechanics.

18. Short Summary of the Elements of Band Spectra

Band spectra are emitted by molecules. These spectra also consist of lines but the lines lie so close to each other that in a spectroscope of not too high resolution they merge into bands (this gave the name to these spectra). In these spectra the number of lines is extremely large.

The theory of band spectra is complicated by the fact that the energy of a molecule contains not only the energy of the electrons but it includes the oscillation energy due to the oscillatory motion of the atoms of the molecule, as well as the rotational energy produced by the rotation of the molecule around some definite axes.

A diatomic molecule can be schematized by a dumb-bell model. In such a molecule the two atoms, or rather the nuclei, can vibrate, oscillate with respect to each other; moreover, the molecule can rotate around the axis perpendicular to the straight line joining the two atoms. The molecule possesses besides the rotation and oscillation energy, electron energy as well.

The rotation energy can be easily calculated for a diatomic molecule. If the moment of inertia is Θ and the constant angular velocity $\dot{\phi}$, then

$$E = E_{kin} = \frac{1}{2}\Theta\dot{\phi}^2 . \tag{18.1}$$

The momentum canonically conjugated to ϕ is

$$P_\phi = \frac{\partial E_{kin}}{\partial \dot{\phi}} = \Theta\dot{\phi} .$$

43

Applying the quantum condition one obtains

$$\oint p_\phi \, d\phi = 2\pi\Theta\dot{\phi} = mh \,, \qquad (18.2)$$

$$\dot{\phi} = \frac{mh}{2\pi\Theta} \,, \qquad (18.3)$$

where m is an integer, the so-called rotational quantum number. Putting the expression of $\dot{\phi}$ into (18.1) and denoting the energy belonging to m by E_m, one gets

$$E_m = \frac{m^2 h^2}{8\pi^2\Theta} \,. \qquad (18.4)$$

It is worth mentioning that the experimental investigation of the spectra led to the result that in the expression of E_m, $m(m + 1)$ has to be written instead of m^2. Quantum mechanics, as will be shown, leads to the same result.

The oscillation energy can be plotted directly if the vibrations are supposed to be harmonic, because a diatomic molecule is a linear oscillator. If the frequency of oscillation is denoted by v_0, the oscillation energy is

$$E_n = nhv_0 \,, \qquad (18.5)$$

where n is the *oscillation quantum number*. In the present treatise this is regarded as an integer; quantum mechanics will, however, furnish half values. Regarding the spectral lines the results are not affected whether n is integer or a half number.

If the electron energy is denoted by E_{el}, the total energy is

$$E = E_{el} + nhv_0 + \frac{m(m + 1)\, h^2}{8\pi^2\Theta} \,. \qquad (18.6)$$

E_{el} gives the biggest part in E, it is followed by nhv_0, and the rotation energy is the smallest. The frequency of the lines is again received from Bohr's frequency conditions. For n and m the following selection rules hold: $\Delta n = \pm 1$, $\Delta m = \pm 1$. The whole theory is much more complicated than that of the line spectra. Here only the basis of the theory could be shown, the theory itself could not be treated fully.

CHAPTER 3

The Elements of Wave Mechanics

19. General Survey

Although Bohr's theory could account for numerous phenomena, it still required considerable modifications. Bohr's theory is in several aspects unsatisfactory and it even contains some inconsistencies.

For instance, Bohr's theory could not explain why the electron revolving around the nucleus does not radiate. The introduction of the quantum conditions is unexplainable in general. In Bohr's theory the atomic problems are discussed in terms of classical mechanics and the quantum conditions are only forced upon them, but cannot be explained or deduced. The allowed electron orbits are selected on the basis of these quantum conditions but a more profound reasoning is lacking. One of the discrepancies is that the orbital quantum number l had to be introduced instead of the azimuthal one k, in order to stay in accordance with experiments. A similar contradiction occurs in the theory of diatomic molecules — as has already been mentioned — that instead of m^2 (the square of the rotational quantum number) $m(m + 1)$ had to be inserted into the rotational energy term to be in agreement with experiments. In the case of the Zeeman effect a similar problem arises: $j(j + 1)$ has to be written instead of j^2. Finally we only mention that Bohr's theory leads already for the simplest many-body problem, the He-atom, to incorrect energy values.

All this called for amendments and modifications of Bohr's theory. Quantum mechanics proved to furnish these additions, eliminating the problems mentioned and surpassing Bohr's theory considerably. Quantum mechanics originated from two seemingly quite different theories: Heisenberg's matrix mechanics and Schrödinger's wave mechanics. It soon turned out that the two theories, founded approximately at the same time, are two different forms of the same theory. Wave mechanics is much more vividly descriptive than matrix mechanics and is more suitable for the solution of concrete problems. Matrix mechanics is highly abstract but indispensable for the profound understanding of quantum mechanics. Here we are concerned only with wave mechanics.

20. The Heisenberg Principle

First the so-called Heisenberg principle will be dealt with. This is of capital importance in quantum mechanics.

According to classical mechanics (the mechanics preceding quantum mechanics) measurement accuracy had no limitations in principle. In practice, of course, the imperfections of the measuring instruments sets limits to the accuracy of measurements, but it seemed obvious that by improving the instrumentation the accuracy could be improved infinitely. This is valid in quantum mechanics, too, if only one variable is measured (in a system). It was Heisenberg who recognized that two canonically conjugate variables of a system cannot be determined simultaneously with arbitrary accuracy. The reason for this is the following: the measurement of one of the variables necessarily brings about an interference in the system and this changes the value of the other (canonically conjugate) variable.

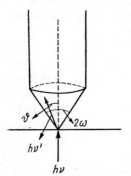

FIG. 8. Heisenberg's gedanken-experiment

Thus, for example, during the measurement of a coordinate of a particle, the particle receives a small momentum and this deteriorates the measuring accuracy of the simultaneous momentum measurement.

For getting an insight into Heisenberg's principle let us perform the following gedanken-experiment. Let us imagine an electron mounted on a microscopic slide and moving with a certain velocity. The electron should be irradiated with monochromatic light of frequency v. The observation consists of the following process: the light, or rather if we look at the elementary process, a photon, deviates from its original direction due to the interaction with the electron, in short the Compton effect occurs. In dealing with the Compton effect it can be proved that due to the scattering the momentum of the deviated photon (more exactly the component perpendicular to the direction of the impinging light-ray) changes by $(hv'/c)\sin\vartheta$, where ϑ is the angle between the original and the final direction of the photon. In first approximation $(hv/c)\sin\vartheta$ can be written instead of this expression, substituting v for v'. As seen from Fig. 8, only those (deviated) photons reach the microscope that lie within the cone of angle 2ω. Thus, in our case, δ, the angle of deviation, has to lie between $\pm\omega$ in order that the deviated photon should get into the microscope. That means that the momentum change of the photons reaching the microscope has to lie between $- (h\nu/c)\sin \omega$ and $+ (h\nu/c)\sin \omega$, thus, after the Compton effect the momentum uncertainty is in the order of $\Delta p = (h\nu/c) \sin \omega$. Taking into consideration that $c/\nu = \lambda$,

$$\Delta p \approx \frac{h}{\lambda} \sin \omega . \qquad (20.1)$$

Let us turn now to the problem of the position determination. A position determination performed with light of wavelength λ is coupled with an uncertainty of the order of λ, namely the dimensions of the diffraction image are of the order of λ. Moreover, the dimensions of the diffraction image depend on the aperture (2ω) of the microscope as well, giving a factor proportional to $1/\sin \omega$. Thus the uncertainty in the position determination (Δq) is of the order of $\lambda/\sin \omega$, i.e.

$$\Delta q \approx \frac{\lambda}{\sin \omega} . \qquad (20.2)$$

That means that a higher accuracy is reached in the position determination if very short wavelengths and large apertures are used, but this would increase Δp.

If the uncertainties of the simultaneous position and momentum determinations are multiplied with each other:

$$\Delta q \Delta p \approx h . \qquad (20.3)$$

Heisenberg's relation is essentially expressed by this equation. Its more accurate formulation is, however, as follows: if p and q are canonically conjugate variables, and if Δq and Δp are the theoretical limitations occurring during the simultaneous measurements of q and p, then

$$\Delta q \Delta p \geqq h . \qquad (20.4)$$

It is necessary to emphasize that the Heisenberg principle relates to the product of Δq and Δp; in principle one of them can be arbitrarily small if the other is increased at a rate demanded by the relation. Thus, the Heisenberg relation does not contain any predictions regarding the accuracy of the measurement of a single variable; this can be increased − if the technical difficulties are disregarded − without limitations. If one of Δq and Δp is zero, i.e. if an absolutely accurate determination of one of the variables were possible, the uncertainty of the other variable at the same instance would be infinitely large, that is to say, nothing would be known about it.

Heisenberg's principle demands considerable restrictions only in atomic dimensions due to the smallness of h. From the relation it follows that

$$\Delta v_x \geqq \frac{h}{m_0} \frac{1}{\Delta x}, \tag{20.5}$$

where Δv_x is the uncertainty of the x-component of the velocity.

Let us first regard a macroscopic example. Consider a ball with 1 g mass. Its place should be determined with an accuracy of 10^{-4} cm. This is a very good accuracy in macroscopic measurements. In this case

$$\Delta v_x \geqq \frac{6.63 \times 10^{-27}}{1} \times \frac{1}{10^{-4}} \text{ cm s}^{-1} = 6.63 \times 10^{-23} \text{ cm s}^{-1}. \tag{20.6}$$

This is an extremely small limit of error which surpasses by far our demands as to the accuracy of the velocity.

In the case of atomic dimensions the situation is quite different. Let us suppose that the position and velocity of an electron have to be determined simultaneously. The uncertainty limit of determining the place of the electron cannot be taken greater by any means than 10^{-8} cm if the electron has to be investigated in the atom. Inserting the electron mass and $\Delta q = 10^{-8}$ cm into the above-mentioned uncertainty relation, the uncertainty of the velocity of the electron is

$$\Delta v_x \geqq \frac{6.63 \times 10^{-27}}{9.11 \times 10^{-28}} \times \frac{1}{10^{-8}} \text{ cm s}^{-1} = 7.3 \times 10^8 \text{ cm s}^{-1}. \tag{20.7}$$

The smallest value of Δv_x is roughly the three-fold of the value determined for the velocity of an electron revolving in the innermost Bohr orbit of the H-atom. That means that the uncertainty limit of the velocity is very high.

Therefore the electron orbit cannot be followed along the route ascribed to it in Bohr's theory. The reason for this is that if the position of an electron is determined within the limits of atomic dimensions, the electron is ejected from its orbit due to the Compton effect. Therefore it is impossible in principle to observe the Bohr orbits.

Heisenberg's relation is one of the most important concepts of theoretical physics. In classical physics it was postulated that all the characteristic values of a mechanical state can be determined with arbitrary accuracy at a given time. In reality however, there are limits to the accuracy: it is not possible to determine with arbitrarily great accuracy every characteristic simultaneously. It is in principle impossible to determine with absolute accuracy all the characteristic mechanical data of a system simultaneously;

48

thus, for example, the three coordinates and the velocity of a point mass at a given initial time. That is why it is in principle impossible to determine from the data received by an investigation made at the initial time the data of the system at a later time. Thus in the new mechanics equations or inter-relations have to be disregarded where from given exact initial conditions the data of the system could be determined at a later moment. The question is not "what follows from a precisely defined initial state?" but the question has to be modified: what follows from a state which is inaccurately defined in the classical sense? It is obvious that this question cannot be answered so accurately as in classical mechanics, only probability predictions can be given for the later state of the system. Even from this it can be seen that the new mechanics, the quantum mechanics, is a *statistical theory*. Of course, in the theory the characteristic values of a system at a certain instant of time, may be regarded uncertain only to such a degree as demanded, due to theoretical reasons, by the Heisenberg relation; the experimental errors occurring due to technical limitations are, naturally, not included in this uncertainty.

Coming back to Heisenberg's relation, the following has still to be emphasized. As the relation holds for every pair of canonically conjugate variables, it is valid also for the energy and the time (these are also canonically conjugate). From the Hamiltonian equations, i.e. $dq/dt = \partial H/\partial p$, if we put $p = H = E$, it follows that $dq/dt = 1$, i.e. $q = t$. Thus Heisenberg's relation holds also for E and t:

$$\Delta E \Delta t \geqq h . \tag{20.8}$$

That means that if the quantum energy is known with absolute exactness, nothing can be said about the time the system spends in this state.

This relation is the cause why no theory can deal with the process of the emission of a photon as a process developing in time. Such a photon-emission theory ought to be expected to deliver the time dependence of the energy during the process of radiation emission. A relation like this, however, could not be checked, as can be easily seen from the relation. On the other hand, in the case of macroscopic systems, e.g. the radiation by an antenna, no difficulties arise due to the relation, for the energy and time uncertainty can be neglected because of the smallness of h.

Heisenberg's relation showed quite new ways for the theory; namely it became obvious that it cannot be the aim of the theory to account for the inner processes of the atoms, unobservable because of theoretical causes, but its task can only be to construct a mechanism suitable to deduce observable quantities, thus, for instance, the wavelength of the spectral lines or the energy levels of the atom from such a model where only the elementary building blocks and the forces acting between them are known, but the integration constants of classical mechanics are unknown, as they cannot be determined all at the same time.

21. Causality in Quantum Mechanics

In connection with Heisenberg's relation the question arises how the principle of causality has to be formulated in quantum mechanics. Before approaching this question the classical mechanical form of this principle will be briefly introduced. There the causality can be formulated in the following form: in the case of a closed system always the same states result if the initial conditions are the same. In quantum mechanics causality does not hold in this form, namely, if an experiment is performed with an atomic system along these lines, that is to say, if we investigate how an atomic system develops when always starting from the same initial state within the limitations of the theory, then it will be observed that the development is generally different in every case. It could be argued that this does not contradict the above-mentioned classical formulation of causality, as we always set off from different initial conditions, and the only cause for not being able to distinguish between them is the theoretical limitation regarding the accuracy of observation, and thus processes corresponding to different initial conditions are actually observed. If it were possible to perform observations beyond the theoretical limitations, a situation where causality is valid in the classical formulation would probably be found. Such argument and speculation is, however, aimless. It cannot be the task of a physicist to determine and investigate what the situation would be if different unrealizable conditions were realized. The physicist focuses his attention on the real observable universe; he can deal only with things that are really observable. From this it follows that if it is absolutely impossible to distinguish between two states, they have to be regarded as the same. That means that in the new mechanics the above-mentioned formulation of causality cannot be accepted.

In this respect the question arises how it is possible that in classical mechanics, in spite of the theoretical limitations of the measuring accuracy, from a given initial condition the later states of the system can be determined unambiguously. This can be explained by the fact that classical mechanics deals with macroscopic bodies and for these the theoretical limit of measuring accuracy is, due to the smallness of h, so small — as already shown by an example — that it is of no importance at all. That means that the principle of causality in its classical form does not hold in the macroscopic world either, only this cannot be observed, as the measuring instruments due to technical limitations, do not furnish us with sufficiently accurate data. However, if Planck's constant were larger, also a macroscopic point mass would change its velocity when illuminated for a position determination.

The next question is, how the principle of causality has to be formulated in quantum mechanics. As already mentioned, it is not possible to derive in the exact classical form from the initial conditions of an atomic system its later state, because the initial data are not known with full accuracy. However, they are known well enough for determining the probability that the result of an observation performed later will be a given state of the system. Without mathematics the quantum mechanical principle of causality can be formulated as follows: a given initial condition does not determine the future development of the system, but it determines the probability that the development will take place in a definite manner. That means that in quantum mechanics one may derive later states only in such a statistical sense from a given state of the system.

22. The Phase and Group Velocity of Waves

In the following some concepts of wave motion (necessary in the later explanations) will be introduced. Let us first consider a simple periodic wave, ϕ, written in a complex form. If the x-axis of the coordinate system is taken along the direction of wave propagation,

$$\phi = A e^{i(\omega t - kx)}, \tag{22.1}$$

where ω is the angular frequency, k the wave number. The amplitude of the wave is denoted by A, the time by t. If T is the time of one complete period and λ is the wavelength

$$\omega = \frac{2\pi}{T} \quad \text{and} \quad k = \frac{2\pi}{\lambda}. \tag{22.2}$$

The velocity of the wave, u is obtained from

$$\omega t - kx = \text{const}. \tag{22.3}$$

by total derivation with respect to t in the following form:

$$u = \frac{dx}{dt} = \frac{\omega}{k} = v\lambda. \tag{22.4}$$

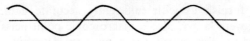

FIG. 9. Simple periodic wave

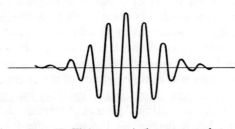

FIG. 10. Finite wave train; wave packet

The simple periodic wave produced like this is always infinite, that means that along the axis, disregarding the momentary nodes, there is always and everywhere a wave motion, the maximum value of which is the amplitude A (see Fig. 9).

If a wave has to be created having a measurable amplitude only in a given interval, and beyond this interval it should be everywhere equal to zero, then waves of different ω and k have to be superposed. Due to interference these will strengthen each other in definite places producing a measurable amplitude, in other places they will weaken each other, and by selecting the waves appropriately the amplitude will be zero. To produce a shorter wave packet, waves with broader ω and k distribution have to be superposed (see Fig. 10).

The expression of the so-called wave packet produced by this superposition is

$$\Phi = \int_{k_0-\eta}^{k_0+\eta} A(k) \, e^{i[\omega(k)t-kx]} \, dk \,. \tag{22.5}$$

If the $(k_0 - \eta, k_0 + \eta)$ interval is small enough, the result can be regarded as a wave with a changing amplitude differing only a bit from the original one. The exponent can be written in the following form:

$$\omega t - kx = \omega_0 t - k_0 x + (\omega - \omega_0)t - (k - k_0)x \,. \tag{22.6}$$

Substituting this into the expression of the wave packet:

$$\Phi = C e^{i(\omega_0 t - k_0 x)}, \tag{22.7}$$

where

$$C = \int_{k_0-\eta}^{k_0+\eta} A(k) \, e^{i[(\omega-\omega_0)t-(k-k_0)x]} \, dk \,. \tag{22.8}$$

The velocity of the wave packet is given by the velocity of the place where C is constant. C is constant where

$$(\omega - \omega_0)t - (k - k_0)x = \text{const} \,. \tag{22.9}$$

And thus the velocity of the wave packet, i.e. the group velocity, is

$$u_g = \frac{dx}{dt} = \frac{\omega - \omega_0}{k - k_0} = \frac{d\omega}{dk} , \qquad (22.10)$$

supposing that $\omega - \omega_0$ and $k - k_0$ are small enough, that is to say, η is small. The $d\omega/dk$ differential quotient refers to the value $k = k_0$. Earlier the following expression was received for u:

$$u = \frac{\omega}{k} , \qquad (22.11)$$

from this

$$k = \frac{\omega}{u} . \qquad (22.12)$$

This yields

$$\frac{1}{u_g} = \frac{d}{d\omega}\left(\frac{\omega}{u}\right) = \frac{d}{dv}\left(\frac{v}{u}\right) , \qquad (22.13)$$

where $v = \omega/2\pi$ is the frequency. This is one of the well-known forms of the group velocity expression. From this it can be seen that the group velocity equals the phase velocity unless the latter is a function of v, i.e. unless a dispersion exists.

The group velocity is a very important concept and, as we have seen, marks the propagation velocity of the wave packet, while the phase velocity is the velocity of the single sine waves composing the group. It is not possible to transmit signals with the valleys and hills of an infinite sine wave. Signalling (acoustically or electromagnetically) can be performed only by means of wave packets. This can be seen from the simple fact that during signalling a wave motion has to be started or stopped and thus the signalling wave cannot be an infinitely long sine wave. From this it follows that the group velocity is equal to the so-called signal velocity.

23. Matter Waves

In Sections 6 and 7 the dual nature of light as well as of the electron and atomic rays has been introduced. Their corpuscular and wave characteristics were demonstrated. By means of these waves (especially in the case of light) one may also deduce Heisenberg's relation that shows the very close connection existing between the corpuscules and the so-called *matter waves* belonging to them. The concept of matter waves originates from de Broglie; in the following these waves will be discussed in detail.

In the case of light the paths of the photons, that is the light rays are the orthogonal trajectories of the wave fronts. If the openings on the optical screens are wide compared with the wave length of the light, that is, if geometrical optics is valid, diffraction can be neglected and the paths of the light rays can be determined by Fermat's principle:

$$\int_{P_0}^{P} n\,ds = \text{extremum}, \qquad (23.1)$$

where P_0 and P are two points along the path of the light ray, n is the refractive index that has to be known as function of position, ds is the path element. Thus Fermat's principle means the following: forming $\int_{P_0}^{P} n\,ds$ along the real path of the light ray and comparing this with the integral formed along other paths connecting P_0 and P, the former is an extremum. Fermat's principle can be written also in the following form:

$$\delta \int_{P_0}^{P} n\,ds = 0, \qquad (23.2)$$

where the path between P_0 and P has to be varied, but the starting and end points remain unchanged. If the reciprocal proportionality between the refractive index and the phase velocity u of the light waves is taken into consideration ($n = c/u$, where c is the light velocity in vacuum), Fermat's principle can be written also in the following form:

$$\delta \int_{P_0}^{P} \frac{ds}{u} = 0. \qquad (23.3)$$

Fermat's principle is valid only in geometrical optics, but if the order of magnitude of the holes in the optical screens is equal to the wavelength of light and diffraction takes place, geometrical optics becomes invalid and so the occurring phenomena cannot be described any more by light rays. When describing the phenomena one has to start from the wave nature of light, that is from wave theory.

In classical mechanics the path of a point mass can be determined from Hamilton's principle which is similar to Fermat's principle serving the determination of the path of the light rays. This similarity becomes specially apparent if only such paths are compared where the variation of the total energy E is everywhere equal to zero. In this case a new principle, called Maupertuis' principle, can be deduced from Hamilton's principle. Hamilton's principle states the following:

$$\delta \int_{t_0}^{t} L\,dt = \delta \int_{t_0}^{t} (2T - E)\,dt = 0 , \tag{23.4}$$

where $L = T - V$, and T is the kinetic, V the potential energy and t is the time. If the total energy is not varied, then it gives

$$\delta \int_{t_0}^{t} 2T\,dt = 0 . \tag{23.5}$$

In the case of a point mass with mass m_0 and velocity v this can be written in the following form:

$$\delta \int_{t_0}^{t} m_0 v^2 dt = \delta \int_{t_0}^{t} m_0 v \frac{ds}{dt}\,dt = \delta \int_{P_0}^{P} m_0 v\,ds = 0 , \tag{23.6}$$

where ds is the path element. If we take now the constancy of m_0 into consideration, we gain Maupertuis' principle according to which:

$$\delta \int_{P_0}^{P} v\,ds = 0 . \tag{23.7}$$

If we would like to assign waves to the corpuscles moving according to the laws of mechanics, similarly to light waves belonging to photons passing along light rays, and if we would like Fermat's principle to be valid also for these waves, then, as seen from a comparison of the formulae (23.3) and (23.7), the phase velocity (u) of these waves has to be proportional to $1/v$. In the case of a potential energy V the velocity is

$$v = \sqrt{\frac{2}{m_0} (E - V)} , \tag{23.8}$$

and thus for u the following relation has to hold

$$u = \frac{B}{\sqrt{E - V}} , \tag{23.9}$$

where B is a proportionality constant.

It has to be emphasized that the waves associated with the corpuscles are not electromagnetic waves. The associated waves are of electromagnetic character only in the case of photons.

We still must give the frequency v of these so-called de Broglie waves. It seems to be obvious to determine this by the following equation:

$$E = hv ,\qquad(23.10)$$

whence

$$u = \frac{B}{\sqrt{hv - V}} .\qquad(23.11)$$

That means that for these associated waves, as u is a function of v, a dispersion holds.

The hypothesis of de Broglie should stand here, because of its basic importance, as put originally (*Journal de Physique*, vol. 1, p. 1, 1924): "Whenever a material corpuscle in a given coordinate system possesses in the most general sense an energy E in this coordinate system a periodic phenomenon is present too with a frequency v defined by the equation $E = hv$ (h is Planck's constant)." We only mention that de Broglie regards the photons as corpuscles. By introducing Planck's constant into the equation defining the frequency of the waves de Broglie established the connection between the waves and quantum theory.

In the following we would like to deal in detail with this relationship between waves and quantum theory. We saw that a dispersion exists for the de Broglie waves, and thus — as discussed in the previous Section — a distinction must be drawn between the phase velocity u of the wave and its group velocity (u_g). It was demonstrated as well that the group velocity is equal to the signal velocity, i.e. equal to that velocity by which, for example, signals of electromagnetic origin propagate; thus this is the velocity that can be measured by electromagnetic registering apparatus. The velocity of a point mass can, of course, also be measured, so that the group velocity of the de Broglie waves associated with the corpuscles has to be equal to the velocity of the corpuscles (the point mass), in our case with v. For the group velocity, we have seen that the following formula holds:

$$\frac{1}{u_g} = \frac{d}{dv}\left(\frac{v}{u}\right).\qquad(23.12)$$

As u_g has to be equal to v, we get

$$\frac{d}{dv}\left(\frac{v}{u}\right) = \frac{1}{v} = \frac{1}{\sqrt{\frac{2}{m_0}(E - V)}} = \frac{1}{\sqrt{\frac{2}{m_0}(hv - V)}},\qquad(23.13)$$

where hv has been written instead of E. After integration

$$\frac{v}{u} = \int \frac{dv}{\sqrt{\dfrac{2}{m_0}(hv - V)}} = \frac{1}{h}\sqrt{2m_0(hv - V)}, \qquad (23.14)$$

so that

$$u = \frac{hv}{\sqrt{2m_0}}\frac{1}{\sqrt{hv - V}}. \qquad (23.15)$$

Thus if the constant B is made equal to $hv/\sqrt{2m_0}$ a dispersion law is reached, that satisfies on the one hand the requirement that $1/u$ be proportional with v, and on the other hand that $u_g = v$. It has to be mentioned that the constant B can contain hv, i.e. E, because in the case of Maupertuis' principle the paths to be compared belong to the same E.

The phase velocity of the de Broglie wave in a field with potential V is:

$$u = \frac{E}{\sqrt{2m_0(E - V)}} = \frac{hv}{\sqrt{2m_0(hv - V)}}. \qquad (23.16)$$

From this the following expression is obtained for the wavelength, in the case of a field again with a potential V:

$$\lambda = \frac{u}{v} = \frac{h}{\sqrt{2m_0(hv - V)}} = \frac{h}{\sqrt{2m_0(E - V)}}. \qquad (23.17)$$

Thus in a field, where $V = 0$, the wavelength is:

$$\lambda_0 = \frac{h}{\sqrt{2m_0 E}} = \frac{h}{m_0 v} = \frac{h}{p}, \qquad (23.18)$$

as in this case $E = \frac{1}{2}m_0 v^2$. If the waves come from a field with $V = 0$ into a field with a potential V, this corresponds to the case when the waves come into a field where the index of refraction relative to the field of zero potential is

$$n = \frac{\lambda_0}{\lambda} = \sqrt{\frac{E - V}{E}}. \qquad (23.19)$$

As n is a function of position, the space with potential V which is a function of position, is equivalent to a region with a changing index of refraction, where the orthogonal trajectories of the wave fronts, that is the paths of the corpuscles are generally curved.

If experience, i.e. the experimental results did not refer to the fact that it is really necessary to associate the corpuscles with the waves, the former discussion would be a vain and superfluous theoretical speculation. Experience shows, however, that diffraction can be produced by electron and atom rays (see Section 7), i.e. these rays show besides their corpuscular nature a wave nature, too. The wavelength is in the case of $V = 0$, as we saw, equal to the wavelength of the de Broglie wave, that is to h/p. It has also been mentioned that according to experiments light also shows a dual character, namely a photon and a wave character. The physical contents of the discussions regarding the de Broglie waves is given by these experimental facts.

24. The Schrödinger Equation

In optics, when diffraction is discussed, the starting-point must be the wave equation. Similarly in atomic mechanics, where we deal with phenomena where the distances are of the order of the wavelength of the de Broglie waves, for discussing these phenomena a wave equation of the de Broglie waves has to be used, similar to the optical wave equation. This is the following one:

$$\nabla^2 \Psi = \frac{1}{u^2} \frac{\partial^2 \Psi}{\partial t^2} ; \qquad u = \frac{E}{\sqrt{2m_0(E - V)}}. \tag{24.1}$$

In this equation the meaning of Ψ is still uncertain. At the present we suppose only that the square of Ψ, and in the case of complex Ψ the square of its absolute value, i.e. $\Psi\Psi^*$ (where Ψ^* is the conjugate complex of Ψ) is the extent of the intensity of the corpuscular beam, for instance, an electron beam, quite similarly to optics where in the solution of the wave equation the square of the wave vector, or of its component yields the value of the light intensity.

As in wave theory generally, Ψ is also here supposed to be a function, exactly periodic in time:

$$\Psi = \psi \times e^{-2\pi i \nu t} = \psi \times e^{-2\pi i (E/h)t} \tag{24.2}$$

where ψ is still a function of the space coordinates. Substituting (24.2) into (24.1) for ψ the fundamental Schrödinger equation, the basis of the whole of atomic mechanics, is reached:

$$\nabla^2 \psi + \frac{8\pi^2 m_0}{h^2} (E - V) \psi = 0. \tag{24.3}$$

This equation can be obtained from the energy expression

$$H = \frac{1}{2m_0}(p_x^2 + p_y^2 + p_z^2) + V(x, y, z), \qquad (24.4)$$

by the following formal substitutions. We write operators: Instead of the components of the momentum,

$$p_x \longrightarrow \frac{h}{2\pi i} \frac{\partial}{\partial x}, \quad p_y \longrightarrow \frac{h}{2\pi i} \frac{\partial}{\partial y}, \quad p_z \longrightarrow \frac{h}{2\pi i} \frac{\partial}{\partial z}. \quad (24.5)$$

The following operators are assigned to the squares of the momentum components:

$$p_x^2 \longrightarrow -\frac{h^2}{4\pi^2} \frac{\partial^2}{\partial x^2}, \quad p_y^2 \longrightarrow -\frac{h^2}{4\pi^2} \frac{\partial^2}{\partial y^2}, \quad p_z^2 \longrightarrow -\frac{h^2}{4\pi^2} \frac{\partial^2}{\partial z^2};$$
$$(24.6)$$

these can be regarded formally as the squares of the above-mentioned operators. Substituting these into the expression of the energy, the so-called Hamilton or energy operator is gained, applying this on a ψ function, i.e. multiplying formally the energy operator from the right-hand side with function ψ and making this expression equal with $E\psi$, the Schrödinger equation is obtained:

$$H\left(x, y, z, \frac{h}{2\pi i} \frac{\partial}{\partial x}, \frac{h}{2\pi i} \frac{\partial}{\partial y}, \frac{h}{2\pi i} \frac{\partial}{\partial z}\right) = E\psi,$$

$$-\frac{1}{2m_0}\left(\frac{h^2}{4\pi^2} \frac{\partial^2\psi}{\partial x^2} + \frac{h^2}{4\pi^2} \frac{\partial^2\psi}{\partial y^2} + \frac{h^2}{4\pi^2} \frac{\partial^2\psi}{\partial z^2}\right) + V\psi = E\psi,$$

$$-\frac{h^2}{8\pi^2 m_0}\left(\frac{\partial^2\psi}{\partial x^2} + \frac{\partial^2\psi}{\partial y^2} + \frac{\partial^2\psi}{\partial z^2}\right) + V\psi = E\psi,$$

$$\nabla^2\psi + \frac{8\pi^2 m_0}{h^2}(E - V)\psi = 0.$$

It has to be mentioned that here to V, which is a function of the coordinates, an operator of multiplication with V has been assigned and from this it follows that to the coordinates the operator of multiplication with coordinates is assigned.

At the present moment all this is a mere formalism, but this formalism already enables us to state the Schrödinger equation in the case of, for instance, generalized coordinates. If q_i and p_i are the position coordinates and the corresponding conjugate momenta, then the following operators correspond to the conjugate momenta p_i and p_i^2:

$$p_i \longrightarrow \frac{h}{2\pi i} \frac{\partial}{\partial q_i}, \qquad p_i^2 \longrightarrow -\frac{h^2}{4\pi^2} \frac{\partial^2}{\partial q_i^2}. \qquad (24.7)$$

These substitutions may, however, be performed only after the adequate symmetrization of the Hamilton function. In this case the Schrödinger equation will have formally the following form

$$H\left(q_i, \frac{h}{2\pi i} \frac{\partial}{\partial q_i}\right) \psi = E\psi. \qquad (24.8)$$

The energy E can be eliminated from the Schrödinger equation, namely

$$\Psi = \psi \times e^{-(2\pi i/h)Et}. \qquad (24.9)$$

Thus

$$\frac{\partial \Psi}{\partial t} = -\frac{2\pi i}{h} E\Psi \quad \text{and} \quad \frac{\partial^2 \Psi}{\partial t^2} = -\frac{4\pi^2}{h^2} E^2 \Psi. \qquad (24.10)$$

Substituting the value of u from the second equation of (24.1) into the first one

$$\nabla^2 \Psi = 2m_0(E - V)\frac{1}{E^2} \frac{\partial^2 \Psi}{\partial t^2}. \qquad (24.11)$$

From the second equation of (24.10)

$$\frac{1}{E^2} \frac{\partial^2 \Psi}{\partial t^2}$$

can be expressed and this expression can be put into (24.11); hence

$$\nabla^2 \Psi = 2m_0(E - V)\left(-\frac{4\pi^2}{h^2} \Psi\right) = -\frac{8\pi^2 m_0}{h^2} E\Psi + \frac{8\pi^2 m}{h^2} V\Psi. \qquad (24.12)$$

Eliminating $E\Psi$ by means of the first equation of (24.1)and rearranging the equation we have:

$$\nabla^2 \Psi - \frac{8\pi^2 m_0}{h^2} V\Psi + \frac{4\pi i m_0}{h} \frac{\partial \Psi}{\partial t} \quad 0. \qquad (24.13)$$

This equation is called the time-dependent Schrödinger equation. This equation is necessary for the discussion of problems where V contains a time variable, i.e. when V is a function of time. In this case the function Ψ is naturally also a function of time.

The time-dependent Schrödinger equation can be produced, for example, in the case of generalized coordinates in the following simple way. Equation (24.8) holds, naturally, not only for Ψ but also for ψ. This can be seen immediately if the equation is multiplied from the right-hand side (or from the left-hand side as the H operator does not act on the time) by $e^{-(2\pi i/h)Et}$. Thus

$$H\left(q_i, \frac{h}{2\pi i}\frac{\partial}{\partial q_i}\right)\Psi = E\Psi . \tag{24.14}$$

Eliminating $E\Psi$ by the help of the first equation of (24.10), the time-dependent Schrödinger equation is obtained for the case of generalized coordinates

$$H\left(q_i, \frac{h}{2\pi i}\frac{\partial}{\partial q_i}\right)\Psi = -\frac{h}{2\pi i}\frac{\partial\Psi}{\partial t} . \tag{24.15}$$

An analogy exists between Schrödinger's equation and the Hamilton–Jacobi equation of classical mechanics. If H does not contain the time explicitly, then the so-called shortened Hamilton–Jacobi equation has the following form:

$$H\left(q_i, \frac{\partial S_0}{\partial q_i}\right) = E , \tag{24.16}$$

where

$$p_i = \frac{\partial S_0}{\partial q_i} \tag{24.17}$$

and S_0 is the shortened Hamilton *principal* function.

By comparing (24.16) with (24.8), the time-independent Schrödinger equation, the analogy becomes apparent. At the same time some basic differences become evident as well. Whereas in case of the Hamilton–Jacobi equation $(\partial S_0/\partial q_i)^2$ is to be substituted in place of p_i^2 in the Hamiltonian expression, in producing Schrödinger's equation $(h/2\pi i)^2\ \partial^2/\partial q_i^2$ has to be written in place of p_i^2. It can be shown that the Hamilton–Jacobi equation follows from the Schrödinger equation when $h \to 0$, as generally if $h \to 0$ the quantum theory passes over to classical mechanics. The *unshortened Hamilton–Jacobi* equation is the following one:

$$H\left(q_i, \frac{\partial S}{\partial q_i}\right) = -\frac{\partial S}{\partial t} , \tag{24.18}$$

where $\partial S/\partial q_i = p_i$ and this equation shows the same analogy with (24.15) as the one existing between (24.16) and (24.18). Thus $(h/2\pi i)(\partial \Psi/\partial t)$ corresponds to $\partial S/\partial t$. In this case S is Hamilton's principal function.

25. The Derivation of the Schrödinger Equation from a Variational Principle

The basic equation of wave mechanics, the Schrödinger equation, can be derived, similarly to the equations of motion of classical mechanics, from a variational principle. The equations of motion of classical mechanics are the Euler–Lagrange differential equations of the Hamilton principle, the Schrödinger equation is a Euler–Lagrange differential equation of a variational principle the form of which is, in the case of a system consisting of N particles, the following:

$$\delta \int L d\tau = 0 , \tag{25.1}$$

where

$$L = \sum_{i=1}^{N} \frac{h^2}{8\pi^2 m_i} \left(\frac{\partial \psi^*}{\partial x_i} \frac{\partial \psi}{\partial x_i} + \frac{\partial \psi^*}{\partial y_i} \frac{\partial \psi}{\partial y_i} + \frac{\partial \psi^*}{\partial z_i} \frac{\partial \psi}{\partial z_i} \right) + U\psi\psi^* . \tag{25.2}$$

The function ψ has to satisfy the following normalizing condition:

$$\int \psi^* \psi \, d\tau = 1 . \tag{25.3}$$

In eq. (25.2) m_i is the mass of particle i, U is the potential energy of the entire system; $d\tau$ is a space element of the $3N$ dimensional space.

Condition (25.3) can be taken into consideration by introducing a Lagrange multiplier $-E$ and thus the variational principle takes the following form:

$$\delta \int (L - E\psi^*\psi) \, d\tau = 0 . \tag{25.4}$$

Performing the variation we get the Schrödinger equation for a system of N particles

$$\sum_{i=1}^{N} \frac{h^2}{8\pi^2 m_i} \nabla_i^2 \psi + (E - U) \psi = 0 , \tag{25.5}$$

where ∇_i^2 is the Laplace operator relating to particle i. From this it can be seen that E, i.e. -1 times the Lagrange multiplier, is equal to the energy parameter of the system.

By using the Hamilton operator H eq. (25.5) can be written in the following form:

$$H\psi = E\psi . \tag{25.6}$$

From this the following expression is gained for the energy parameter E:

$$E = \frac{\int \psi^* H \psi \, d\tau}{\int \psi^* \psi \, d\tau}.$$ (25.7)

Through partial integration we get

$$\int L \, d\tau = \int \psi^* H \psi \, d\tau .$$ (25.8)

Using this equation for E, the following expression is reached from (25.7):

$$E = \frac{\int L \, d\tau}{\int \psi^* \psi \, d\tau} .$$ (25.9)

From this it follows, that if the variational principle is taken into consideration, the eigenvalues of the Schrödinger equation are the extrema of expression (25.9), and it can be shown that they are its minima.

From (25.8) it follows that the variational principle can be formulated in another way, too:

$$\delta \int \psi^* (H - E) \, \psi \, d\tau = 0 .$$ (25.10)

The energy minima of eq. (25.7) are thus equal to the eigenvalues of (25.6) or to those of (25.5), as these equations are identical.

26. Eigenfunctions, Eigenvalues

The Ψ function that satisfies the Schrödinger equation has a similar connection with the corpuscles as light waves have with the photons. Thus no corpuscles will reach those points where $\Psi\Psi^*$ or Ψ vanishes. If we look for a Ψ solution of the Schrödinger equation that corresponds to an electron bound to an atom, then this Ψ has to vanish at infinity, as no bound electron can get as far as that. Furthermore, a physically meaningful solution of the Schrödinger equation is expected to be everywhere continuous and a single-valued function of space. It was a great achievement of quantum mechanics in contrast to the old quantum theory that these quite obvious requirements were enough for determining the discrete energy levels of atomic structures. In the former theory foreign quantum conditions were necessary for obtaining these levels.

An interesting consequence of the continuity and single-valuedness of the Ψ function is the fulfilment of the old Bohr quantum conditions for sufficiently large closed orbits. According to (23.17) the wavelength of the wave

Ψ describing the movement of a point mass m_0 is

$$\lambda = \frac{h}{\sqrt{2m_0(E - V)}}.$$ (26.1)

This generally changes along the orbit from point to point. Let us consider an element dq of the orbit, and let us suppose that even more complete waves fall on this orbit element. The orbit can be decomposed into such parts only if the change of the wavelength along the orbit is slow enough, that is, if the orbit is large enough. The number of waves falling on a section dq is dq/λ; on the entire orbit $\oint dq/\lambda$ waves fall, where the integration refers to the entire closed orbit. The single-valuedness of Ψ requires that the wavelength could be measured an integral times onto the orbit (see Fig. 11). Thus

$$\oint \frac{dq}{\lambda} = \oint \frac{1}{h} \sqrt{2m_0(E - V)} \, dq = n$$ (26.2)

where n is an integer. From the expression of the energy, $E = p^2/2m_0 + V$ we get the following formula for p:

$$p = \sqrt{2m_0(E - V)}.$$ (26.3)

From this we get from (26.2)

$$\oint p \, dq = nh.$$ (26.4)

Thus the Bohr quantum condition is obtained.

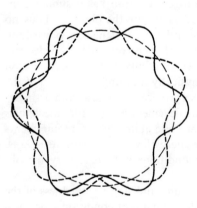

FIG. 11. de Broglie waves along a circular orbit

It has to be emphasized once more that these considerations are valid only if the orbits are large enough and are closed. In other cases the results received by the help of wave mechanics and those received by the classical Bohr theory differ from each other, thus the two theories are not equivalent. In every case when the results of the two theories are different, experience always decides for wave mechanics.

It follows from the theory of linear differential equations that the boundary conditions on Ψ and the requirements that Ψ has to be a continuous and single-valued function of space are fulfilled only in the case of certain parameters E. Thus the energy parameter E cannot take arbitrary values. Those E values for which the Schrödinger equation, with the above-mentioned conditions, can be solved, are called the eigenvalues of the equation. These are written

$$E_1, E_2, \ldots, E_i, \ldots$$

The solutions belonging to these eigenvalues are written

$$\Psi_1, \Psi_2, \ldots, \Psi_i, \ldots$$

and are called the eigenfunctions of the equation. We are interested primarily in the eigenvalues, and their determination will be our main aim. The determination of the eigenfunctions is, at the present moment, of secondary importance. The best way to become aquainted with the methods of wave mechanics and the solution of the Schrödinger equation is the investigation of some special cases, and we will come back to some of them in due time. Before this, however, some questions of a general character have to be dealt with.

First the physical meaning of the function Ψ will be introduced. Originally Schrödinger interpreted the function Ψ in the following way. He supposed that the electron belonging to an atom is not an electrically charged point mass with a definite mass, but the mass and charge of the electron is continuously distributed (smeared out) in the atom, and the charge density of it is proportional to $\Psi\Psi^*$. This original interpretation of Schrödinger had to be revised, because in the case of many-electron problems Ψ is a function of $3N$ coordinates, where N is the number of the electrons, and thus Ψ cannot be interpreted in the normal three-dimensional space. Ψ is a function of the $3N$ dimensional so-called configuration space, and Schrödinger's interpretation of the Ψ-function cannot be retained. Nowadays the Ψ-function is interpreted statistically, after Born. According to this $\Psi\Psi^*dv$ gives the probability that the electron is present in the element of space dv. In the case of more electrons dv means the element of configuration space, and the above-mentioned expression gives the probability that the electrons are present in the element dv of configuration space. The basic difference between this and Schrödinger's interpretation is that in this case the electron is henceforward regarded as an electrically charged point mass. The experimental findings are always gained either as the average value of the investigation of many atoms or as the result of prolonged measurements, especially if electrons bound to atoms are investigated, and thus although Ψ has been interpreted statistically, one may speak about the charge density of the electron, or electrons. In this case the spatial average or the average over a long period (long according to the time measure of the atoms, in reality it is a very short time interval) is meant.

Let us now consider a system being in the kth quantum state. We shall deal with the eigenvalue of such a system, for instance, with that of an electron bound to an atom. The energy of the system, in our case of the electron in this state should be denoted by E_k, and the eigenfunction is:

$$\Psi_k = e^{-(2\pi i/h)E_k t} \cdot \psi_k(x, y, z). \tag{26.5}$$

We see that

$$\Psi_k \Psi_k^* = \psi_k \psi_k^* . \tag{26.6}$$

The ψ_k functions contain an arbitrary factor, because the Schrödinger equation is homogeneous and thus its solutions can contain an arbitrary multiplying constant. This constant can be chosen in such a form that the following relation should be valid:

$$\int \psi_k \psi_k^* \, dv = 1 . \tag{26.7}$$

The integration has to be extended over the whole space. If ψ_k (and Ψ_k) satisfy this condition, they are called normalized. The physical meaning of this is the following: $\psi_k \psi_k^* \, dv$ is the probability that an electron being in the kth quantum state is present in the space element dv; as the electron can certainly be found somewhere in space, the total probability for this is unity, thus expression (26.7) must necessarily hold.

Let us now prove the following important relation: if the kth and lth eigenvalues are different, then the following relation holds for the corresponding eigenfunctions:

$$\int \psi_k \psi_l^* \, dv = 0 , \tag{26.8}$$

where the integration has to be extended over the whole of space. This relation means that the eigenfunctions belonging to different eigenvalues are orthogonal.

This statement can be proved as follows: both ψ_k and ψ_l are solutions of the Schrödinger equation; thus,

$$\nabla^2 \psi_k + \frac{8\pi^2 m_0}{h^2} (E_k - V) \psi_k = 0 \tag{26.9}$$

and

$$\nabla^2 \psi_l^* + \frac{8\pi^2 m_0}{h^2} (E_l - V) \psi_l^* = 0 . \tag{26.10}$$

If the first equation is multiplied by ψ_l^*, the second one by ψ_k, and from this latter equation the former one is subtracted, we get

$$\psi_k \nabla^2 \psi_l^* - \psi_l^* \nabla^2 \psi_k = \frac{8\pi^2 m_0}{h^2} (E_k - E_l)\, \psi_k \psi_l^* . \qquad (26.11)$$

Multiplying this equation by dv and integrating it over the whole of space, Green's theorem can be applied to the left-hand side of the equation. According to this

$$\int (\psi_k \nabla^2 \psi_l^* - \psi_l^* \nabla^2 \psi_k)\, dv = \int \left(\psi_k \frac{\partial \psi_l^*}{\partial n} - \psi_l^* \frac{\partial \psi_k}{\partial n} \right) df . \qquad (26.12)$$

The right-hand side integral is a surface integral, it has to be extended over the surface that bounds the volume over which the left-hand side integral has been extended. The right-hand side integrand contains differentiations with respect to n, the mean differentiations according to the normal of the surface showing outwards. As will be seen in connection with special problems, ψ vanishes exponentially at infinity. If the left-hand side integral is extended over the whole of space, and accordingly the right-hand side integral is extended over a surface infinitely far, the right-hand side integral vanishes, thus we have:

$$\int (\psi_k \nabla^2 \psi_l^* - \psi_l^* \nabla^2 \psi_k)\, dv = 0 , \qquad (26.13)$$

if the integration is extended over the whole of space.

Using (26.11), for $E_k \neq E_l$, we obtain

$$\int \psi_k \psi_l^*\, dv = 0 , \qquad (26.14)$$

if the integration is extended over the whole of space. Thus the orthogonality relation has been proved.

Regarding the solution of the Schrödinger equation the following has to be mentioned. As the equation is linear and homogeneous, if $\Psi_1, \Psi_2, \ldots, \Psi_n$ are solutions, then the following expression is also a solution:

$$\Psi = c_1 \Psi_1 + c_2 \Psi_2 + \ldots + c_n \Psi_n , \qquad (26.15)$$

where the c_i coefficients are constants. The physical meaning of these coefficients is obviously the following: $|c_i|^2$ denotes the probability that the system, e.g. an electron bound to an atom, is in the ith quantum state.

67

We postulate for coefficients c_i that

$$\sum_i |c_i|^2 = 1 .\tag{26.16}$$

This, of course, can always be obtained. If the electron is, for example, in the kth state, then all the c_i except c_k are zero and c_k equals 1. Hence

$$\Psi_k = e^{-(2\pi i/h)\,E_k t}\,\psi_k .$$

This means that a state with a constant energy E_k, a so-called stationary state, exists.

27. The Quantum-mechanical Interpretation of the Quantities of Classical Mechanics, the Operator Concept

First the quantum-mechanical equivalents of the various quantities of classical mechanics (e.g. coordinates, momenta, etc.) have to be introduced. It will be reasonable to keep a concrete problem in view; let us deal with an electron bound to an atom, the corresponding eigenfunction should be Ψ, which is supposed to be normalized to unity. Thus $\Psi\Psi^* dv$ is the probability that the electron is present within dv. Let us deal first with a single co-ordinate, its classical mechanical equivalent should be q. In quantum mechanics the q-coordinate of an electron of an atom can be determined by measurements in the following way. The q-coordinate of this electron (bound to the atom) is measured a great many times (and after each measurement the electron has to be brought back into its original quantum state), and the average value of these different measurement results has to be calculated. Accordingly in quantum mechanics the expectation value of the q-coordinate measurement is gained by producing the average value of q with the $\Psi\Psi^*$ probability distribution, thus

$$\bar{q} = \int q\Psi\Psi^* dv = \int \Psi^* q\,\Psi dv .\tag{27.1}$$

$\bar{q}$, i.e. the quantum-mechanical average of q is the $\bar{q}$ value attainable (or expected) as the average of q values received from measurements (Erwartungswert); it is called the quantum-mechanical average of q.

In order to determine the quantum-mechanical average of the momentum and the components of momentum, respectively, first the quantum-mechanical expression of the current will be deduced. For this reason we start from the time-dependent Schrödinger equation, the form of which, if, for in-

68

stance, a single corpuscle of mass m_0 is regarded, is the following:

$$\nabla^2 \Psi - \frac{8\pi^2 m_0}{h^2} V\Psi + \frac{4\pi i m_0}{h} \frac{\partial \Psi}{\partial t} = 0 . \tag{27.2}$$

The conjugate complex of this equation is naturally valid too

$$\nabla^2 \Psi* - \frac{8\pi^2 m_0}{h^2} V\Psi* - \frac{4\pi i m_0}{h} \frac{\partial \Psi*}{\partial t} = 0 . \tag{27.3}$$

Multiplying the first equation by $\Psi*$ and the second equation by Ψ and subtracting from the second one the first one, we get the following equation:

$$-\frac{4\pi i m_0}{h} \left(\Psi* \frac{\partial \Psi}{\partial t} + \Psi \frac{\partial \Psi*}{\partial t} \right) = -\frac{4\pi i m_0}{h} \frac{\partial(\Psi\Psi*)}{\partial t} = \Psi*\nabla^2\Psi - \Psi\nabla^2\Psi* . \tag{27.4}$$

As it is obvious that

$$\frac{\partial}{\partial x} \left(\Psi* \frac{\partial \Psi}{\partial x} - \Psi \frac{\partial \Psi*}{\partial x} \right) = \Psi* \frac{\partial^2 \Psi}{\partial x^2} - \Psi \frac{\partial^2 \Psi*}{\partial x^2} ,$$

and similar relations hold for y and z as well, thus

$$\Psi*\nabla^2\Psi - \Psi\nabla^2\Psi* = \operatorname{div} (\Psi* \operatorname{grad} \Psi - \Psi \operatorname{grad} \Psi*) . \tag{27.5}$$

So that

$$-\frac{\partial(\Psi*\Psi)}{\partial t} = \frac{h}{4\pi i m_0} \operatorname{div} (\Psi* \operatorname{grad} \Psi - \Psi \operatorname{grad} \Psi*) . \tag{27.6}$$

If we multiply this equation by e, the charge of the corpuscle, then

$$-\frac{\partial(\Psi\Psi*e)}{\partial t} = \frac{he}{4\pi i m_0} \operatorname{div} (\Psi* \operatorname{grad} \Psi - \Psi \operatorname{grad} \Psi*) . \tag{27.7}$$

Let this equation be compared with the continuity equation of electro-dynamics, according to which

$$-\frac{\partial \rho}{\partial t} = \operatorname{div} \mathbf{i} , \tag{27.8}$$

where ρ is the charge density and $\mathbf{i}$ the current density. This equation is the analogy of the corresponding equation of hydrodynamics and expresses that in a given volume the electric charge can decrease only if it flows away from there. Generally this equation expresses the theorem of the conservation of electricity. It can be seen from a comparison of (27.7) and (27.8) that in quantum mechanics the current density has to be defined by the following equation:

$$\mathbf{i} = \frac{he}{4\pi i m_0} (\Psi^* \text{ grad } \Psi - \Psi \text{ grad } \Psi^*). \tag{27.9}$$

$\mathbf{s} = (h/4\pi i m_0) (\Psi^* \text{ grad } \Psi - \Psi \text{ grad } \Psi^*)$, is thus the probability that the corpuscle passes perpendicularly through a unit surface area in unit time. From this it follows that the quantum-mechanical expression of the current $\mathbf{I}$ is:

$$\mathbf{I} = e\mathbf{v} = \frac{he}{4\pi i m_0} \int (\Psi^* \text{ grad } \Psi - \Psi \text{ grad } \Psi^*) \, dv. \tag{27.10}$$

This yields the quantum-mechanical value of the momentum of the corpuscle:

$$\mathbf{p} = \frac{h}{4\pi i} \int (\Psi^* \text{ grad } \Psi - \Psi \text{ grad } \Psi^*) \, dv. \tag{27.11}$$

By partial integration

$$\int \Psi \text{ grad } \Psi^* dv = - \int \Psi^* \text{ grad } \Psi \, dv. \tag{27.12}$$

This can be demonstrated in the easiest way by proving it for the single components, that is, by showing that

$$\int \Psi \frac{\partial \Psi^*}{\partial x} \, dv = - \int \Psi^* \frac{\partial \Psi}{\partial x} \, dv,$$

and that similar equations hold for the y and z components. Their correctness is evident. Now

$$\mathbf{p} = \frac{h}{2\pi i} \int \Psi^* \text{ grad } \Psi \, dv, \tag{27.13}$$

and hence it follows that

$$\bar{p}_x = \frac{h}{2\pi i} \int \Psi^* \frac{\partial \Psi}{\partial x} \, dv, \qquad \bar{p}_y = \frac{h}{2\pi i} \int \Psi^* \frac{\partial \Psi}{\partial y} \, dv,$$

$$\bar{p}_z = \frac{h}{2\pi i} \int \Psi^* \frac{\partial \Psi}{\partial z} \, dv. \tag{27.14}$$

A given component of the momentum, e.g. the z component, is obtained by applying the $(h/2\pi i)(\partial/\partial z)$ operator to the Ψ function, and by multiplying this function $(h/2\pi i)(\partial\Psi/\partial z)$ by Ψ^* and integrating it over the whole of space. The formula obtained for $\bar{p}_z$ can be formally regarded as the mean value of the operator $(h/2\pi i)(\partial/\partial z)$ with Ψ and Ψ^*, produced by multiplying formally the operator from the right side with Ψ, from the left side with Ψ^* and integrating the product. Thus it can be said that in quantum mechanics p_z is represented by the operator $(h/2\pi i)(\partial/\partial z)$. Generally the p_i coordinate of the momentum canonically conjugate to q_i is represented (or produced) in quantum mechanics by the operator $(h/2\pi i)(\partial/\partial q_i)$. This gives a more profound foundation for the introduction of this operator in the formal deduction of the Schrödinger equation. It has to be mentioned that the value of $\bar{p}_z$ determined by the above integral, that is the quantum-mechanical value of p_z, provides that value of p_z that is gained as the average value of the measurements of p_z, in case the system is in a quantum state corresponding to Ψ.

The quantum-mechanical value of q obtained earlier can be interpreted by analogy with the quantum-mechanical value of the momentum components so that the operator of the multiplication by q is applied to Ψ, and this function $(q\Psi)$ is multiplied by Ψ^* and integrated over the whole of space. This gives a more profound interpretation of the formal operation of the multiplication by q.

Thus, generally, to all physical quantities operators can be assigned, and for every physical quantity its quantum-mechanical average can be determined.

To the potential energy, V, as it is the function of the space coordinates, the operator of multiplying by V is assigned, thus the quantum-mechanical value of V is:

$$\bar{V} = \int \Psi^* V \Psi \, dv . \tag{27.15}$$

According to the above-mentioned, the operator belonging to the kinetic energy, in the case of a point mass m_0, is

$$\frac{1}{2m_0}\left(\frac{h}{2\pi i}\operatorname{grad}\right)^2 = -\frac{h^2}{8\pi^2 m_0}\left(\frac{\partial^2}{\partial x^2} + \frac{\partial^2}{\partial y^2} + \frac{\partial^2}{\partial z^2}\right) = -\frac{h^2}{8\pi^2 m_0}\nabla^2. \tag{27.16}$$

And thus the quantum-mechanical value of the kinetic energy is:

$$\bar{T} = -\frac{h^2}{8\pi^2 m_0}\int \Psi^* \nabla^2 \Psi \, dv . \tag{27.17}$$

We have already seen that in the case of generalized coordinates the so-called Hamilton-operator representing the energy is:

$$H\left(q_i, \frac{h}{2\pi i} \frac{\partial}{\partial q_i}\right).$$

In the case of generalized coordinates that gives the expression for the energy in quantum mechanics:

$$\bar{E} = \int \Psi^* H\left(q_i, \frac{h}{2\pi i} \frac{\partial}{\partial q_i}\right) \Psi dv. \qquad (27.18)$$

In the case of a point mass m_0 and Cartesian coordinates this expression takes the form

$$\bar{E} = \int \Psi^* \left(- \frac{h^2}{8\pi^2 m_0} \nabla^2 + V\right) \Psi dv =$$

$$- \frac{h^2}{8\pi^2 m_0} \int \Psi^* \nabla^2 \Psi dv + \int \Psi^* V \, \Psi dv. \qquad (27.19)$$

Furthermore, we saw that $\Psi = \psi e^{-(2\pi i/h)Et}$, and this implies

$$- \frac{h}{2\pi i} \frac{\partial \Psi}{\partial t} = E\Psi, \qquad (27.20)$$

so that the $-(h/2\pi i)(\partial/\partial t)$ operator represents the energy. This operator contains differentiation with respect to t, whereas the Hamilton operator contains only the coordinates and differentiations with respect to them. It is seen from the above that

$$\bar{E} = - \frac{h}{2\pi i} \int \Psi^* \frac{\partial \Psi}{\partial t} dv. \qquad (27.21)$$

The following operator represents the components of angular momentum for example the component z is

$$N_z \longrightarrow \frac{h}{2\pi i} \left(x \frac{\partial}{\partial y} - y \frac{\partial}{\partial x}\right). \qquad (27.22)$$

The square of the absolute value of the angular momentum is represented by the operator:

72

$$N^2 \longrightarrow -\frac{h^2}{4\pi^2} \sum_{x,y,z} \left(y\frac{\partial}{\partial z} - z\frac{\partial}{\partial y}\right)\left(y\frac{\partial}{\partial z} - z\frac{\partial}{\partial y}\right)$$

$$= -\frac{h^2}{4\pi^2}\left[\left(y\frac{\partial}{\partial z} - z\frac{\partial}{\partial y}\right)\left(y\frac{\partial}{\partial z} - z\frac{\partial}{\partial y}\right) + \left(z\frac{\partial}{\partial x} - x\frac{\partial}{\partial z}\right)\left(z\frac{\partial}{\partial x} - x\frac{\partial}{\partial z}\right)\right.$$

$$\left. + \left(x\frac{\partial}{\partial y} - y\frac{\partial}{\partial x}\right)\left(x\frac{\partial}{\partial y} - y\frac{\partial}{\partial x}\right)\right] = -\frac{h^2}{4\pi^2}\left[(y^2+z^2)\frac{\partial^2}{\partial x^2} + (z^2+x^2)\frac{\partial^2}{\partial y^2}\right.$$

$$+ (x^2 + y^2)\frac{\partial^2}{\partial z^2} - 2yz\frac{\partial^2}{\partial y\partial z} - 2zx\frac{\partial^2}{\partial z\partial x} - 2xy\frac{\partial^2}{\partial x\partial y}$$

$$\left. - 2x\frac{\partial}{\partial x} - 2y\frac{\partial}{\partial y} - 2z\frac{\partial}{\partial z}\right]. \tag{27.23}$$

The quantum-mechanical values of these operators can be produced similarly to those mentioned above.

Generally, if a function $L(q_i, p_i)$ is given, then the operator associated to it is $L[q_i, (h/2\pi i)(\partial/\partial q_i)]$ and the corresponding quantum-mechanical value is

$$\bar{L} = \int \Psi^* L\left(q_i, \frac{h}{2\pi i}\frac{\partial}{\partial q_i}\right)\Psi\,dv. \tag{27.24}$$

Normally the same relations hold for the quantum-mechanical values as for the classical mechanical ones, thus, for example, $\bar{p}_z = m\dot{\bar{z}}$.

On the basis of the former, one may associate an operator to every physical quantity. It was shown that the Schrödinger equation can be established by applying the Hamilton operator to a Ψ function and putting the expression obtained equal to $E\Psi$, that is to a multiple of Ψ. The equation gives the possible values of E. Applying the operator H to any function another function is gained, which is usually not a multiple of the original Ψ. The solutions of the Schrödinger equation are just those functions, which yield a multiple (E) of the original function if the Hamilton operator is applied to them. The multiplication factors are the eigenvalues of the operators and of the equations, the corresponding functions are the eigenfunctions. The eigenvalues of the operator are the physical quantities corresponding to the operators, in the present case the possible values of the energy. This statement can be generalized as for every physical quantity there exists an operator representing it. The eigenfunctions of the operators are those functions for which applying the operator they turn into their

λ-times value. The λ are the eigenvalues of the operator giving the possible values of the physical quantity corresponding to the operator.

For example, the operator belonging to the coordinate q is the operator of multiplication by q, the eigenvalue is q itself:

$$q\phi = q\phi \, .$$

As can be seen q can take any value. The eigenfunctions, the ϕ, are arbitrary as well.

The eigenfunctions and eigenvalues (identified here by α) of the x-component of the momentum can be determined from the following equation:

$$\frac{h}{2\pi i} \, \frac{\partial \chi}{\partial x} = \alpha \chi \, .$$

The solution of this equation is $\chi = \chi_0 \, e^{(2\pi i/h)p_x x}$, where χ_0 is an arbitrary function of y and z. p_x is obtained as eigenvalue.

All these are essential for the elucidation and more profound foundation of quantum mechanics.

28. The Quantum Theory of the Electron Spin

In the preceding section the laws by which the quantum-mechanical operators corresponding to the classical mechanical quantities can be produced have been introduced. In microphysics there are also quantities which have no counterpart in classical mechanics. Such a quantity is, for example, the electron spin. The operators of these quantities cannot be given on the basis of a general rule, but one has to find by trial and error an operator with consequences being in agreement with experiment.

The experimental facts in connection with the electron spin have already been discussed. According to this, the component in any direction of space of the spin S of the electron can take only the values $+h/4\pi$ or $-h/4\pi$. The magnetic moment connected with the spin is equal to $-(e/m_0 c)S$, the component of this in any direction of space can be only $-eh/4\pi m_0 c$ or $+eh/4\pi m_0 c$. It is usual to relate the spin state of the electron to the z-axis of the coordinate system and to introduce a spin coordinate s for describing the spin state. The coordinate s can take only the values $s = \pm 1$, which correspond to the $\pm h/4\pi$ values of the spin z-components. The wave function of the electron, if the phenomena related to the spin have to be described too, depends besides the Cartesian coordinates x, y, z also on the spin coordinate:

$$\Psi = \Psi(x, y, z, s, t).\tag{28.1}$$

The spin coordinate s dependence of the Ψ wave function can be given, for example, by the following little table:

$s =$	$+1$	-1
$\Psi(x, y, z, s, t) =$	$\Psi_+(x, y, z, t)$	$\Psi_-(x, y, z, t)$

Ψ_+ and Ψ_- depend only on the space and time coordinates. The wave function (28.1) can be identified unambiguously by the Ψ_+ and Ψ_- functions. It is usual to write this pair of functions in the form of a 1×2 matrix

$$\Psi = \begin{pmatrix} \Psi_+(x, y, z, t) \\ \Psi_-(x, y, z, t) \end{pmatrix}.\tag{28.2}$$

This wave function is called a *(Pauli) spinor*. Ψ_+ and Ψ_- are the components of the spinor. It is important to mention that any spinor can be written as the linear combination of two special spinors, namely of

$$\chi_{+\frac{1}{2}} = \begin{pmatrix} 1 \\ 0 \end{pmatrix} \quad \text{and} \quad \chi_{-\frac{1}{2}} = \begin{pmatrix} 0 \\ 1 \end{pmatrix}.\tag{28.3}$$

The form of the wave function (28.2) in terms of these spinors is

$$\Psi = \Psi_+(x, y, z, t)\chi_{+\frac{1}{2}} + \Psi_-(x, y, z, t)\chi_{-\frac{1}{2}}.\tag{28.4}$$

Three methods for describing the spin dependence have been introduced: the expressions (28.1), (28.2) and (28.4).

These three methods are absolutely equivalent and always the most suitable one is to be chosen when investigating concrete problems.

Now we would like to deal with the quantum-mechanical operators applied to the wave function Ψ. Let us start from the spinor form (28.2) of the wave function. The most general linear operator through which from a spinor with two components another one can be produced is a 2×2 matrix. The quantum-mechanical operators can also be written in the form of such matrices. The effect of the operator

$$A = \begin{pmatrix} A_{11} & A_{12} \\ A_{21} & A_{22} \end{pmatrix}\tag{28.5}$$

on the wave function (28.2) is obtained by matrix multiplication:

$$A\Psi = \begin{pmatrix} A_{11} & A_{12} \\ A_{21} & A_{22} \end{pmatrix} \begin{pmatrix} \Psi_+ \\ \Psi_- \end{pmatrix} = \begin{pmatrix} A_{11}\Psi_+ + A_{12}\Psi_- \\ A_{21}\Psi_+ + A_{22}\Psi_- \end{pmatrix}. \tag{28.6}$$

Here the components A_{11}, A_{12}, A_{21} and A_{22} are operators applied only to the space coordinates.

The following matrices can be introduced after Pauli:

$$\sigma_x = \begin{pmatrix} 0 & 1 \\ 1 & 0 \end{pmatrix}, \quad \sigma_y = \begin{pmatrix} 0 & -i \\ i & 0 \end{pmatrix}, \quad \sigma_z = \begin{pmatrix} 1 & 0 \\ 0 & -1 \end{pmatrix}. \tag{28.7}$$

Every 2×2 matrix can be produced as the linear combination of the unity matrix

$$1 = \begin{pmatrix} 1 & 0 \\ 0 & 1 \end{pmatrix}$$

and the Pauli matrices:

$$A = \begin{pmatrix} A_{11} & A_{12} \\ A_{21} & A_{22} \end{pmatrix} = A_0 1 + A_x \sigma_x + A_y \sigma_y + A_z \sigma_z, \tag{28.8}$$

where

$$\begin{aligned}
A_0 &= \tfrac{1}{2}(A_{11} + A_{22}), \\
A_x &= \tfrac{1}{2}(A_{12} + A_{21}), \\
A_y &= \tfrac{1}{2}(A_{12} - A_{21}), \\
A_z &= \tfrac{1}{2}(A_{11} - A_{22}).
\end{aligned} \tag{28.9}$$

If the three Pauli matrices are regarded as the three Cartesian components of a vector operator:

$$\boldsymbol{\sigma} = (\sigma_x, \sigma_y, \sigma_z), \tag{28.10}$$

and if from the components A_x, A_y, A_z of expressions (28.9) an **A** vector is produced, then expression (28.8) can be written in a condensed form as follows

$$A = A_0 + \mathbf{A}\boldsymbol{\sigma}. \tag{28.11}$$

This form of the operator A [or its equivalent (28.8)] is especially reasonable if the form (28.4) of Ψ is applied. When producing $A\Psi$ only the product of the Pauli matrices and the χ_{m_s} ($m_s = \pm \tfrac{1}{2}$) basic spinors is encountered. These, on the other hand, can be easily calculated from the definitions (28.7) and (28.3); the result of this is seen in Table 1.

TABLE 1. *The Effect of the Pauli Matrices on the Basic Spinors*

	$\chi_{+\frac{1}{2}}$	$\chi_{-\frac{1}{2}}$
σ_x	$\chi_{-\frac{1}{2}}$	$\chi_{+\frac{1}{2}}$
σ_y	$i\chi_{-\frac{1}{2}}$	$-i\chi_{+\frac{1}{2}}$
σ_z	$\chi_{+\frac{1}{2}}$	$-\chi_{-\frac{1}{2}}$

Having dealt with the mathematical framework of the quantum theory of electron spin and in order to reflect upon the way in which it is incorporated in physical problems we have to determine which concrete operators correspond to the physical quantities connected with the spin variable. First of all the spin **S**, the operator of the angular momentum of the electron itself has to be known. We are going to demonstrate that the following operator fulfils all the physical requirements concerning the spin:

$$\mathbf{S} = \frac{h}{4\pi}\,\boldsymbol{\sigma}. \tag{28.12}$$

The last line of Table 1 shows that

$$S_z\chi_{+\frac{1}{2}} = +\frac{h}{4\pi}\,\chi_{+\frac{1}{2}} \quad \text{and} \quad S_z\chi_{-\frac{1}{2}} = -\frac{h}{4\pi}\,\chi_{-\frac{1}{2}}. \tag{28.13}$$

The basic spinors χ_{m_s} are eigenfunctions of the z component of the spin, because if S_z is applied to χ_{m_s} it changes into a multiple of χ_{m_s}. The corresponding eigenvalues are $\pm h/4\pi$. This, on the one hand, agrees with the experimental fact that the spin components can be ony $\pm h/4\pi$ in any direction of space, and on the other hand it shows that the spinor $\chi_{+\frac{1}{2}}$ describes a spin pointing in the $+z$-direction, while the spinor $\chi_{-\frac{1}{2}}$ describes one pointing to the $-z$-direction.

Let us determine the component of the spin in an arbitrary direction of space. An arbitrary direction in space is characterized by the polar angles ϑ and ϕ, the rectangular components of the unit vector pointing into the given direction of space are

$$n_x = \sin\vartheta\cos\phi,$$
$$n_y = \sin\vartheta\sin\phi,$$
$$n_z = \cos\vartheta.$$

The component of the spin $\mathbf{S}$ along the direction of space $\mathbf{n}$ is

$$S_n = \mathbf{n}\mathbf{S} = \frac{h}{4\pi} (n_x \sigma_x + n_y \sigma_y + n_z \sigma_z). \qquad (28.14)$$

Let us determine the eigenvalues and eigenfunctions of the S_n operator. As the first step the effect of S_n on the basic spinors χ_{m_s} will be investigated. Using the rules summed up in Table 1:

$$S_n \chi_{+\frac{1}{2}} = \frac{h}{4\pi} (\cos \vartheta \chi_{+\frac{1}{2}} + \sin \vartheta e^{i\phi} \chi_{-\frac{1}{2}}),$$

$$\qquad (28.15)$$

$$S_n \chi_{-\frac{1}{2}} = \frac{h}{4\pi} (\sin \vartheta e^{-i\phi} \chi_{+\frac{1}{2}} - \cos \vartheta \chi_{-\frac{1}{2}}).$$

The eigenvalue equation for the spin component S_n is

$$S_n \chi = \lambda \chi, \qquad (28.16)$$

where λ is the eigenvalue corresponding to the eigenfunction χ. χ can be taken in the form expressed in terms of the basic spinors: $\chi = a\chi_{+\frac{1}{2}} + b\chi_{-\frac{1}{2}}$. When substituting this into the eigenvalue equation, and applying the rules (28.15), the following equation is reached:

$$\frac{h}{4\pi} [(a \cos \vartheta + b \sin \vartheta e^{-i\phi}) \chi_{+\frac{1}{2}} + (a \sin \vartheta e^{i\phi} - b \cos \vartheta) \chi_{-\frac{1}{2}}]$$

$$= \lambda [a\chi_{+\frac{1}{2}} + b\chi_{-\frac{1}{2}}].$$

This equation can hold only if the coefficients of both $\chi_{+\frac{1}{2}}$ and $\chi_{-\frac{1}{2}}$ are identical on the two sides of the equation. From comparing the coefficients of $\chi_{+\frac{1}{2}}$ we get

$$\left(\frac{h}{4\pi} \cos \vartheta - \lambda \right) a + \frac{h}{4\pi} \sin \vartheta e^{-i\phi} b = 0, \qquad (28.17)$$

and from those of $\chi_{-\frac{1}{2}}$ we get

$$\frac{h}{4\pi} \sin \vartheta e^{i\phi} a - \left(\frac{h}{4\pi} \cos \vartheta + \lambda \right) b = 0. \qquad (28.18)$$

These two equations form simultaneous homogeneous linear equations with a and b unknown quantities. A nontrivial solution exists only if the determinant of the simultaneous equations is equal to zero:

$$D \equiv - \left(\frac{h^2}{16\pi^2} \cos^2 \vartheta - \lambda^2 \right) - \frac{h^2}{16\pi^2} \sin^2 \vartheta = 0 \, .$$

From this we get $\lambda^2 = h^2/16\pi^2$, which means the two eigenvalues are $\lambda = \pm h/4\pi$. Thus the spin operator (28.12) agrees with the basic experimental finding that the spin component in an arbitrary direction of space can take only the values $+h/4\pi$ or $-h/4\pi$. By solving eqs (28.17) and (28.18) for the unknown a and b, the eigenfunctions of the spin component S_n can be determined, too. In the case of the eigenvalue $\lambda = +h/4\pi$, the eigenfunction is

$$\chi_\uparrow = \cos \frac{\vartheta}{2} \chi_{+\frac{1}{2}} + \sin \frac{\vartheta}{2} e^{i\phi} \chi_{-\frac{1}{2}} \tag{28.19}$$

while in the case of $\lambda = -h/4\pi$:

$$\chi_\downarrow = - \sin \frac{\vartheta}{2} e^{-i\phi} \chi_{+\frac{1}{2}} + \cos \frac{\vartheta}{2} \chi_{-\frac{1}{2}} \, . \tag{28.20}$$

Table 2 contains the eigenfunctions of the spin components S_x, S_y and S_z.

TABLE 2. *The Eigenfunctions of the Spin Components*

Direction	ϑ	ϕ	$\chi_\uparrow$	$\chi_\downarrow$
x	$\dfrac{\pi}{2}$	0	$\dfrac{1}{\sqrt{2}} (\chi_{+\frac{1}{2}} + \chi_{-\frac{1}{2}})$	$\dfrac{-1}{\sqrt{2}} (\chi_{+\frac{1}{2}} - \chi_{-\frac{1}{2}})$
y	$\dfrac{\pi}{2}$	$\dfrac{\pi}{2}$	$\dfrac{1}{\sqrt{2}} (\chi_{+\frac{1}{2}} + i\chi_{-\frac{1}{2}})$	$\dfrac{1}{\sqrt{2}} (i\chi_{+\frac{1}{2}} + \chi_{-\frac{1}{2}})$
z	0	$-$	$\chi_{+\frac{1}{2}}$	$\chi_{-\frac{1}{2}}$

According to our results the eigenfunctions of the spin components corresponding to different directions in space will be different. This corresponds to the fact that it is impossible to determine simultaneously the spin components in two different directions.

Furthermore, we have to deal with the square of the spin angular momentum **S**, with the spin operator **S²**. Both from Table 1 and from the matrix form of eq. (28.7) it is easy to see that the squares of the Pauli matrices are always equal to unity

$$\sigma_x^2 = \sigma_y^2 = \sigma_z^2 = 1 \, . \tag{28.21}$$

Thus the square of the spin operator is

$$\mathbf{S}^2 = \frac{h^2}{16\pi^2}\,(\sigma_x^2 + \sigma_y^2 + \sigma_z^2) = \frac{3h^2}{16\pi^2}\,. \tag{28.22}$$

That means that the spin operator of (28.12) describes an angular momentum the absolute value of which is $\mathbf{S} = \sqrt{3}h/4\pi$, a universal constant. The absolute value of the electron spin cannot change in any circumstances.

Finally let us introduce the operator of the magnetic moment associated with the spin. As already mentioned this magnetic moment is equal to $-(e/m_0c)\mathbf{S}$, thus

$$\mathbf{\mu}_s = -\frac{e}{m_0 c}\,\mathbf{S} = -\frac{eh}{4\pi m_0 c}\,\mathbf{\sigma}\,. \tag{28.23}$$

The component along the space direction $\mathbf{n}$ of this magnetic moment is $-(e/m_0c)S_n$, and so the eigenvalues of this component are $\pm eh/4\pi m_0 c$, in full accordance with experiment.

29. The Correspondence Principle. Transition Probabilities

We know that the energy levels of the H-atom could already be calculated on the basis of Bohr's theory, and the frequencies of the spectral lines could be determined from the energy levels and from Bohr's frequency condition according to the equation

$$E_i - E_k = h\nu_{ik}\,. \tag{29.1}$$

It was not possible, however, to determine the intensity of the spectral lines, that is, the transition probability for $E_i \rightarrow E_k$, directly from the theory. This lack was supplemented by Bohr's correspondence principle. An exact theoretical foundation of this principle was not possible, but the results following from it showed good agreement with experiments. According to this principle the intensity of the spectral lines corresponding to the transition from the ith quantum state can be determined from the intensity of that radiation that would be emitted by an electron occupying an orbit corresponding to the ith quantum state and emitting according to the classical Maxwell theory. For this electron not only the frequency but also the intensity is exactly defined. This is performed in the Bohr theory in the following way: the angular frequency in the ith quantum orbit shall be denoted by ν. As the coordinates x, y and z of the electron are periodic functions of time, they can according to Fourier be written as a super-position of purely harmonic oscillations; the frequency of these oscillation is an integer multiple of v. Thus, for instance,

$$x = \sum_{n=0}^{\infty} x_n \cos(2\pi n v t + \alpha_n), \qquad (29.2)$$

where the α_n are the phase constants, and the x_n are the amplitudes, determinable by the Fourier theory. Each harmonic oscillation corresponds to a linear harmonic oscillator, and thus, for example, the nth harmonic oscillation corresponds to such a linear harmonic oscillator, i.e. to a vibrating dipole, for which the corresponding absolute value of the electric moment is

$$e x_n \cos(2\pi n v t + \alpha_n). \qquad (29.3)$$

According to the Maxwell theory such an oscillator emits radiation with frequency nv, and the energy emitted per sec is

$$\frac{16\pi^4 e^2}{3c^3} v_n^4 x_n^2$$

where $v_n = nv$ and c is the velocity of light. Taking into account that y and z can be written like x as the superposition of harmonic oscillations, which are also equivalent to oscillators, the following expression is reached for the total energy emitted per sec at frequency v_n:

$$\frac{16\pi^4 e^2}{3c^3} v_n^4 (x_n^2 + y_n^2 + z_n^2). \qquad (29.4)$$

The meaning of y_n and z_n is quite similar to that of x_n. The correspondence principle states the following: if v_n is replaced by the quantum-mechanical frequency

$$v_{i,i-n} = \frac{E_i - E_{i-n}}{h}$$

in expression (29.4), then the expression obtained yields an approximate value of the intensity of the light emitted during the spontaneous transition from the ith quantum state to the i-nth quantum state.

Although this principle yielded only rough values as to the relative intensity of the various lines, the so-called selection rules giving an explanation for the missing of some spectral lines, that is of the forbidden transitions, could be deduced. So it became obvious that in spite of the fact that the theoretical foundation of the correspondence principle was far from being adequate, the principle itself was basically correct.

Let us now consider the problem of how the correspondence principle can be extended into wave mechanics, i.e. of how the correspondence principle can be formulated in wave mechanics in a theoretically consistent form. For this we start from a Ψ state of an atom, e.g. the H-atom which shows in general the following form:

$$\Psi = \sum_i c_i \psi_i e^{-(2\pi i/h) E_i t} . \tag{29.5}$$

Obviously the quantum-mechanical average of x, or in short the quantum-mechanical value of x corresponds to the classical x coordinate of the electron, thus to the expression (29.2). This quantum mechanical value is

$$\bar{x} = \int \Psi x \Psi^* dv = \int x \sum_i c_i \psi_i e^{-(2\pi i/h) E_i t} \cdot \sum_k c_i^* \psi_k^* e^{+(2\pi i/h) E_k t} \, dv$$

$$= \sum_{i,k} c_i c_k^* e^{-(2\pi i/h)(E_i - E_k)t} \cdot \int \psi_k^* x \psi_i \, dv . \tag{29.6}$$

Let us introduce the following notation:

$$\int \psi_k^* x \psi_i \, dv = x_{ki} .$$

That yields

$$\bar{x} = \sum_{i,k} c_i c_k^* x_{ki} \cdot e^{-(2\pi i/h)(E_i - E_k)t} . \tag{29.7}$$

From the definition of x_{ik} it follows that

$$x_{ki} = x_{ik}^* .$$

As the complex conjugate of $c_i c_k^*$ is $c_i^* c_k$, the terms with ik and ki indexes can be combined and thus

$$x = \sum_{i,k} | c_i c_k^* | 2 | x_{ki} | \cos (2\pi v_{ik} t + \alpha_{ik}) , \tag{29.8}$$

where α_{ik} is the sum of the arguments of the complex numbers $c_i c_k^*$ and x_{ki}, whereas v_{ik} can be defined in the following way:

$$v_{ik} = \frac{E_i - E_k}{h} .$$

Naturally similar formulae are valid for y and z.

Hence in wave mechanics, not x, y and z can be produced as the super-position of harmonic oscillations, but $\bar{x}, \bar{y}, \bar{z}$. The different frequencies, however, are not the integer multiples of a given frequency, but the frequencies of the single oscillations are equal to the frequencies of the light waves the atom can emit. These frequencies were obtained in the old theory from Bohr's frequency condition. That means that here full agreement replaces correspondence. It is obvious that in wave mechanics instead of x_n, y_n and z_n the amplitude of the oscillations composing $\bar{x}, \bar{y}$ and $\bar{z}$ have to be written in expression (29.4). The $|c_i c_k^*|$ factors, however, are not to be added to the amplitudes. If these factors were written into the expression (29.4) together with the amplitudes, the following result would be reached. From a state such that all the c, except c_i, are equal to zero, the atom cannot change spontaneously into another quantum state. This, however, contradicts experiments. Thus the $2 | x_{ki} |$, etc., has to be constituted into expression (29.4) for the amplitudes. In wave mechanics the correspondence principle can be formulated in the following way. If an atom is in the ith stationary state, then the probability per sec that it will change spontaneously into a $k < i$ quantum state, is

$$w = \frac{64\pi^4 e^2}{3c^3 h} v_{ik}^3 (| x_{ki} |^2 + | y_{ki} |^2 + | z_{ki} |^2) . \tag{29.9}$$

This expression is not the exact counterpart of (29.4) as the above probability can be obtained from an expression of the intensity giving the emitted energy per sec and corresponding to (29.4) by dividing the latter by the energy hv_{ik} emitted during a single transition. As the emitted energy of a single $i \rightarrow k$ transition is $hv_{ik} = E_i - E_k$, the total light intensity emitted per sec by N atoms being in the ith quantum state is

$$J = N \frac{64\pi^4 e^2}{3c^3} v_{ik}^4 (| x_{ki} |^2 + | y_{ki} |^2 + | z_{ki} |^2) . \tag{29.10}$$

The quantities x_{ki}, y_{ki}, z_{ki} have to be calculated from the normalized eigenfunctions because they are unambiguous only in this case. These important expressions deduced from (29.4) can be derived on the basis of quantum mechanics as well, and there it can even be shown that not only the probability of the emission, but also that of the absorption, that is, the probability of $i \rightarrow k$ transitions for which $k > i$ is proportional to

$$(| x_{ki} |^2 + | y_{ki} |^2 + | z_{ki} |^2) .$$

We must still mention that we could have deduced the expressions w and J perhaps even in a more expressive form if we had started from the quantum-mechanical expectation value of the electric dipole of the atom which is:

$$\mathbf{p} = - \int \Psi^* \mathbf{r} e \, \Psi \, dv = - \sum_{i,k} c_i c_k^* e^{-(2\pi i/h)(E_i - E_k)t} \int \psi_k^* \mathbf{r} e \, \psi_i \, dv , \quad (29.11)$$

where $\mathbf{r}$ is the radius vector drawn from the nucleus to the electron. Since here the dipole moment is time-dependent, the atom emits. The energy emitted per sec is obtained by writing instead of the moment the properly interpreted quantum-mechanical value of the moment into the classical expression. This is a procedure quite similar to that shown above.

Let us draw some further conclusions from the above obtained expressions for w and J. Those cases where for given combinations of i and k; some of the x_{ki}, y_{ki} and z_{ki} expressions vanish are of special interest. If for a given i, k combination

$$x_{ki} = 0 , \quad y_{ki} = 0 , \quad z_{ki} = 0 ,$$

then from this it can be obviously concluded that the $i \rightarrow k$ transition does not take place, i.e. the line corresponding to the ν_{ki} frequency is lacking from the spectrum. This offers a possibility for deducing certain selection rules. If for a given combination only two of the x_{ki}, y_{ki}, z_{ki} values vanish but the third differs from zero, e.g.

$$x_{ki} \neq 0 , \quad y_{ki} = 0 , \quad z_{ki} = 0 ,$$

then the radiation corresponding to the $i \rightarrow k$ transition will be the equivalent of the radiation of a vibrating dipole oscillating in the direction of the x-axis, thus it will be linearly polarized in the direction of the x-axis. If only one of the values x_{ki}, y_{ki}, z_{ki} vanishes, e.g.

$$x_{ki} \neq 0 , \quad y_{ki} \neq 0 , \quad z_{ki} = 0 ,$$

then during the $i \rightarrow k$ transition the atom emits elliptically or circularly polarized light.

On the basis of the above-mentioned, some selection rules may be established for certain systems. Thus, for example, for the linear harmonic oscillator we find that only such transitions are possible where the oscillation quantum number changes by ± 1, that is, the linear harmonic oscillator can emit or absorb at the same time only one quantum of energy $h\nu$. For the H-atom the same selection rules are found, as were already mentioned in connection with the Bohr theory, i.e. the principal and radial quantum numbers (n and n_r, respectively) can change arbitrarily, the orbital quantum number, l only by ± 1, and the magnetic quantum number only by ± 1 or 0.

PART TWO

CHAPTER 4

Simple Examples of Solutions of the Schrödinger Equation

30. Free Mass Point

If a mass point (corpuscle) of m_0 is given, and

$$V = 0, \tag{30.1}$$

then

$$H = T \quad \text{and} \quad T = \frac{1}{2} m_0 v^2, \tag{30.2}$$

where v is the velocity of the mass point.

In this case the form of the Schrödinger equation is the following:

$$\nabla^2 \psi + \frac{8\pi^2 m_0}{h^2} E\psi = 0. \tag{30.3}$$

In case of Cartesian coordinates this equation can be written in a more detailed form

$$\frac{\partial^2 \psi}{\partial x^2} + \frac{\partial^2 \psi}{\partial y^2} + \frac{\partial^2 \psi}{\partial z^2} + \frac{8\pi^2 m_0}{h^2} E\psi = 0. \tag{30.4}$$

The solution of this equation is

$$\psi = A e^{ik(\alpha x + \beta y + \gamma z)}, \tag{30.5}$$

where α, β, γ are direction cosines and A is an arbitrary constant. Their substitution into the equation yields

$$\left[k^2(\alpha^2 + \beta^2 + \gamma^2) - \frac{8\pi^2 m_0}{h^2} E \right] \psi = 0. \tag{30.6}$$

As this has to be valid for every x, y and z, the coefficient of ψ has to vanish. Taking the relation holding for the direction cosines into account we get from here

$$k^2 = \frac{8\pi^2 m_0}{h^2} E, \tag{30.7}$$

that is

$$k = 2\pi \frac{\sqrt{2 m_0 E}}{h} = 2\pi \frac{m_0 v}{h} = 2\pi \frac{p}{h}. \tag{30.8}$$

ψ is a periodic function in the α, β, γ direction, and as seen from the formula of ψ with a period (i.e. wavelength) $2\pi/k$.

$$k = \frac{2\pi}{\lambda}.\tag{30.9}$$

Comparing this with our previous result (30.8) gives

$$\lambda = \frac{h}{p},\tag{30.10}$$

in accordance with the experimental results.

Taking these results into account and the fact that

$$p\alpha = p_x, \qquad p\beta = p_y, \qquad p\gamma = p_z,\tag{30.11}$$

the expression (30.5) of ψ can be transcribed into the following form:

$$\psi = Ae^{(2\pi i/\lambda)(\alpha x+\beta y+\gamma z)} = Ae^{(2\pi i/h)(p_x x+p_y y+p_z z)}\tag{30.12}$$

Ψ can be received from ψ by multiplying the latter by $e^{-2\pi i\nu t}$. Thus

$$\Psi = Ae^{-2\pi i[\nu t-(1/\lambda)(\alpha x+\beta y+\gamma z)]} = Ae^{-(2\pi i/h)[Et-(p_x x+p_y y+p_z z)]}.\tag{30.13}$$

This is a plane wave travelling in the α, β, γ direction, its frequency is $\nu = E/h$, and its wavelength is $\lambda = h/p$.

Let us produce $\psi\psi^*$. We see immediately that

$$\Psi\Psi^* = |A|^2.\tag{30.14}$$

As $|A|^2$ is constant, a quantity independent of position and time, we see that for $V = 0$ the probability that the particle is present at a definite place is everywhere and always the same. Thus nothing is known of the location of the particle at a given time. This is in full accordance with the Heisenberg relation as the momentum coordinates and the energy of the particle appear in our formulae, thus their exact knowledge is supposed, but then nothing can be known about the time the particle spends in this energy state or about the position of the particle.

31. The Plane Rotator

The plane rotator is a mass point revolving on a circular orbit around another fixed point. A mass point revolving around a fixed axis on a circular orbit is, for instance, a plane rotator, as in this case the mass point stays always in the same plane. Our first task is to write the Schrödinger equation; for this reason the energy expression has to be determined. The position of the mass point m_0 on the circular orbit, the radius of which is taken as known, can be described by a simple parameter, namely by the planar polar angle ϕ, i.e. the azimuth.

As $V = 0$, the expression of energy is

$$H = T = \frac{1}{2} \Theta \dot\phi^2 , \tag{31.1}$$

where Θ is the momentum of inertia originating from the rotation around an axis perpendicular to the plane of rotation and through the origin. The momentum coordinate canonically conjugated to ϕ is

$$p_\phi = \frac{\partial T}{\partial \dot\phi} = \Theta \dot\phi . \tag{31.2}$$

Thus

$$H = \frac{1}{2\Theta} p_\phi^2 = E . \tag{31.3}$$

And so the Schrödinger equation is

$$-\frac{1}{2\Theta} \frac{h^2}{4\pi^2} \frac{d^2\psi}{d\phi^2} = E\psi , \tag{31.4}$$

that is

$$\frac{d^2\psi}{d\phi^2} = -\frac{8\pi^2\Theta E}{h^2} \psi . \tag{31.5}$$

The solution of this equation is

$$\psi = Ae^{\pm i \sqrt{(8\pi^2\Theta E/h^2)}\,\phi} , \tag{31.6}$$

where A is an arbitrary constant that can be determined from the normalization conditions.

The eigenvalues of the problem are determined from the requirement that ψ should be a single-valued function of position. In the present case this means that the value of ψ has to be the same at values $\psi = 0$ and

89

$\psi = 2\pi$, from this it follows that in the exponent

$$\sqrt{\frac{8\pi^2 \Theta E}{h^2}}$$

has to be equal to an integer n, thus

$$\sqrt{\frac{8\pi^2 \Theta E}{h^2}} = n , \qquad (31.7)$$

that is

$$E = E_n = \frac{h^2}{8\pi^2 \Theta} n^2 , \quad (n = 0, \pm 1, \pm 2, \pm 3, \ldots) . \qquad (31.8)$$

These E_n values are the energy eigenvalues of the plane rotator.

The plane rotator can take only these values of energy. In this problem we see that quantum mechanics yields the same result as the Bohr theory. In the case of the three-dimensional (rigid) rotator, however, the quantum-mechanical result and that of the Bohr theory are different. This problem, useful in schematizing the rotation of diatomic molecules, will be dealt with shortly.

The eigenfunctions of the plane rotator are

$$\psi = \psi_n = A e^{\pm in\phi} , \qquad (31.9)$$

that is

$$\psi_n = A \cos n\phi , \quad \text{and} \quad \psi_n = A \sin n\phi . \qquad (31.10)$$

This means that two eigenfunctions linearly independent from each other belong to every E_n energy value. Generally if more, for example, z eigenfunctions linearly independent from each other belong to an eigenvalue, the given state is called z times degenerate. Thus in the present case every state is doubly degenerate. The answer to the question why two independent eigenfunctions belong to every state is that in case of a given n the rotator can rotate both in the positive and in the negative direction, and so two states correspond to every n, i.e. to every energy level E_n.

32. The Linear Harmonic Oscillator

The linear harmonic oscillator is a mass point m_0 performing vibrations along a straight line, e.g. along the x-axis, due to a restoring force equal to $-kx$. Thus the potential energy is $V = \frac{1}{2} kx^2$. We know that the frequency

90

of vibration is

$$v = \frac{1}{2\pi} \sqrt{\frac{k}{m_0}} \, . \tag{32.1}$$

The Schrödinger equation shows the following form:

$$\frac{d^2\psi}{dx^2} + \frac{8\pi^2 m_0}{h^2} (E - V) \psi = 0 \, . \tag{32.2}$$

Eliminating k from V, the latter can be written in the following form:

$$V = 2\pi^2 m_0 v^2 x^2 \, . \tag{32.3}$$

Substituting this into the Schrödinger equation we get

$$\frac{d^2\psi}{dx^2} + \left[\frac{8\pi^2 m_0}{h^2} E - \frac{16\pi^4 m_0^2 v^2}{h^2} x^2 \right] \psi = 0 \, . \tag{32.4}$$

Let us introduce the following notation:

$$\lambda = \frac{8\pi^2 m_0}{h^2} E \quad \text{and} \quad \alpha = \frac{4\pi^2 m_0 v}{h} \, . \tag{32.5}$$

By means of these the Schrödinger equation becomes

$$\frac{d^2\psi}{dx^2} + (\lambda - \alpha^2 x^2) \psi = 0. \tag{32.6}$$

Now let us change over from the variable x to a new variable:

$$\xi = \sqrt{\alpha} x \, . \tag{32.7}$$

For that purpose we divide the above equation by α and when taking into consideration that $d\xi = \sqrt{\alpha} \, dx$, we get

$$\frac{d^2\psi}{d\xi^2} + \left(\frac{\lambda}{\alpha} - \xi^2 \right) \psi = 0 \, . \tag{32.8}$$

First the asymptotic solutions of this equation are wanted for large x, that is for large ξ, when the λ/α constant is negligible beside ξ^2. In this case the equation has the form:

$$\frac{d^2\psi}{d\xi^2} = \xi^2 \psi \, . \tag{32.9}$$

91

The asymptotic solution of this equation for large ξ is

$$\psi = e^{\pm \xi^2/2} . \tag{32.10}$$

Namely, if for the sake of simplicity differentiation with respect to ξ is indicated by a prime, we see that

$$\psi' = \pm \xi \psi \quad \text{and} \quad \psi'' = \pm \psi + \xi^2 \psi .$$

Thus for large ξ-values

$$\psi'' \approx \xi^2 \psi .$$

This shows that the expression (32.10) for ψ is really an asymptotic solution of (32.8). As ψ has to vanish at ∞, in the exponent only the negative sign comes into question. Thus the useful asymptotic solution of (32.8) is

$$\psi = e^{-\xi^2/2} . \tag{32.11}$$

We look for a solution of (32.8) in the following form:

$$\psi = e^{-\xi^2/2} u(\xi) . \tag{32.12}$$

In this case we obtain that

$$\psi' = e^{-\xi^2/2} (u' - \xi u) \tag{32.13}$$

and

$$\psi'' = e^{-\xi^2/2} (u'' - 2\xi u' - u + \xi^2 u) . \tag{32.14}$$

Substituting (32.12) and (32.14) into (32.8) and dividing the equation by $e^{-\xi^2/2}$ we find

$$u'' - 2\xi u' + \left(\frac{\lambda}{\alpha} - 1 \right) u = 0 . \tag{32.15}$$

To solve this equation we express u in terms of a power series

$$u = \sum_i a_i \xi^i , \tag{32.16}$$

and substitute this into (32.15), then

$$\sum_i \left[i(i - 1) a_i \xi^{i-2} - 2i a_i \xi^i + \left(\frac{\lambda}{\alpha} - 1 \right) a_i \xi^i \right] = 0 . \tag{32.17}$$

92

Ordering this expression according to the powers of ξ:

$$\sum_i \left[(i + 2)(i + 1) a_{i+2} + \left(\frac{\lambda}{\alpha} - 1 - 2i \right) a_i \right] \xi^i = 0 . \qquad (32.18)$$

This is valid for every value of ξ, if

$$(i + 2)(i + 1) a_{i+2} + \left(\frac{\lambda}{\alpha} - 1 - 2i \right) a_i = 0 . \qquad (32.19)$$

Thus we obtained a two-term recurrence relation for calculating the terms of (32.16). This defines the expression completely, for the first coefficient can be taken as unity.

The eigenfunctions have to vanish at ∞. As ξ tends to ∞, u has to tend to ∞ more slowly, that is in a lower order than (32.11) does. This can be ensured only if in (32.16) i does not tend to ∞, that means that expression (32.16) has to consist of a finite number of terms (it has to be a polynomial). If we wish all the coefficients after $i = n$ to vanish, then writing in formula (32.19) n instead of i the only requirement must be that a_n should vanish, because then all the coefficients $a_{n+2}, a_{n+4}, \ldots$ are to vanish. Let us put

$$\frac{\lambda}{\alpha} - 1 - 2n = 0 , \qquad (32.20)$$

that is

$$\frac{\lambda}{\alpha} = 2n + 1 . \qquad (32.21)$$

When substituting the values of λ and α into this expression we obtain

$$\frac{2E}{h\nu} = 2n + 1 ,$$

that is

$$E = E_n = (2n + 1) \frac{h\nu}{2} = \left(n + \frac{1}{2} \right) h\nu . \qquad (32.22)$$

Thus from the requirement that ψ has to vanish at infinity the energy eigenvalues could be determined. As can be seen E_n differs from its value predicted by the Bohr theory, where $E_n = nh\nu$ was reached. The difference is that, as we can see, in quantum mechanics the energy of the linear oscillator is a half-odd integral multiple of $h\nu$, while in the old quantum theory it was an integral multiple of $h\nu$. This implies a significant difference only when $n = 0$, as in the old quantum theory the energy of the linear oscillator is then zero, while according to quantum mechanics it is $\frac{1}{2} h\nu$. The displacement of the energy levels by $\frac{1}{2} h\nu$ does not produce further important differences compared with Planck's original hypothesis according to which the oscillator can emit or absorb energy in quanta $h\nu$, namely this holds also in quantum mechanics as can be shown by more detailed calculations.

The polynomials (32.16) obtained in the above-mentioned way are called Hermite polynomials, the one belonging to the index n is called a Hermite polynomial of degree n and is designated by $H_n(\xi)$. Although $H_n(\xi)$ is perfectly well defined, it still seems worth giving its differential equation. This can be gained from (32.15) by substituting $\lambda/\alpha = 2n + 1$ which yields

$$H_n''(\xi) - 2\xi H_n'(\xi) + 2nH_n(\xi) = 0 , \qquad (32.23)$$

where n is an integer. Generally this equation defines $H_n(\xi)$, but the polynomials determined from this equation contain an indefinite factor, because the equation is homogeneous. It can be shown that disregarding this indefinite factor, $H_n(\xi)$ is equal to the following expression:

$$H_n(\xi) = (- 1)^n e^{\xi^2} \frac{d^n e^{-\xi^2}}{d\xi^n} . \qquad (32.24)$$

From this

$$H_0 = 1 , \qquad\qquad H_3 = 8\xi^3 - 12\xi ,$$

$$H_1 = 2\xi , \qquad\qquad H_4 = 16\xi^4 - 48\xi^2 + 12 ,$$

$$H_2 = 4\xi^2 - 2 , \qquad . \quad . \quad . \quad . \quad .$$

Consequently the eigenfunction associated with E_n is

$$\psi_n = A_n e^{-\xi^2/2} \cdot H_n(\xi) = A_n(- 1)^n e^{+\xi^2/2} \frac{d^n e^{-\xi^2}}{d\xi^n} , \qquad (32.25)$$

where A_n is a normalizing factor. It can be determined from the equation $\int \psi^* \psi \, dx = 1$ that

$$A_n^2 = \sqrt{\frac{\alpha}{\pi}} \frac{1}{2^n n!} . \qquad (32.26)$$

33. Square Potential Well

Let us investigate the one-dimensional movement (e.g. along the x-axis) of a mass point m_0 in a square potential well. The potential energy is shown in Fig. 12. The origin of the x-axis is in the middle of the potential well; in the range $-a < x < a$ the potential is equal to a constant $-V_0$, otherwise it is zero. The Schrödinger equation becomes:

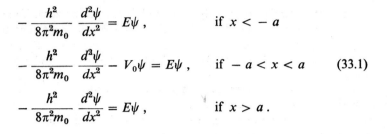

$$-\frac{h^2}{8\pi^2 m_0}\frac{d^2\psi}{dx^2} = E\psi, \qquad \text{if } x < -a$$

$$-\frac{h^2}{8\pi^2 m_0}\frac{d^2\psi}{dx^2} - V_0\psi = E\psi, \qquad \text{if } -a < x < a \qquad (33.1)$$

$$-\frac{h^2}{8\pi^2 m_0}\frac{d^2\psi}{dx^2} = E\psi, \qquad \text{if } x > a.$$

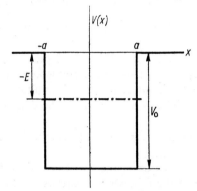

FIG. 12. Square potential well

If the energy E is positive, the kinetic energy of the mass point will be bigger than V_0, and so the mass point can step out of the potential well and move off into infinity. In this case we speak about the scattering of the mass point by the potential well. If $-V_0 < E < 0$, the mass point is bound to the potential well. The present section deals with the bound states, the scattering states will be dealt with in the next section. Let us introduce the following constants:

$$\kappa = \sqrt{-\frac{8\pi^2 m_0}{h^2}E} \quad \text{and} \quad \alpha = \sqrt{\frac{8\pi^2 m_0}{h^2}(E + V_0)}. \qquad (33.2)$$

The general solutions of the differential equations of (33.1) can be written in the following form:

$$\psi(x) = b_1 e^{\kappa x} + c_1 e^{-\kappa x}, \qquad\qquad \text{if} \quad x < -a$$

$$\psi(x) = b_2 \sin(\alpha x) + c_2 \cos(\alpha x), \quad \text{if} \quad -a < x < a \qquad (33.3)$$

$$\psi(x) = b_3 e^{\kappa x} + c_3 e^{-\kappa x}, \qquad\qquad \text{if} \quad x > a.$$

As ψ must vanish as $x \to \pm \infty$, c_1 and b_3 have to be zero. The constants b_1, b_2, c_2 and c_3 can be determined from the requirement that both $\psi(x)$ and $d\psi/dx$ have to be continuous functions at the edges of the well. The conditions of continuity of $\psi(x)$ and $d\psi/dx$ at $x = -a$ and $x = a$ are the following:

$$b_1 e^{-\kappa a} \quad = -b_2 \sin(\alpha a) + c_2 \cos(\alpha a),$$

$$b_1 \kappa\, e^{-\kappa a} = b_2 \alpha \cos(\alpha a) + c_2 \,\alpha \sin(\alpha a), \qquad (33.4)$$

$$c_3 e^{-\kappa a} \quad = b_2 \sin(\alpha a) + c_2 \cos(\alpha a),$$

$$-c_3 \kappa\, e^{-\kappa a} = b_2 \alpha \cos(\alpha a) - c_2 \alpha \sin(\alpha a).$$

A set of simultaneous linear homogeneous equations is obtained; from it the constants b_1, b_2, c_2 and c_3 can be determined. A non-trivial solution is reached only if the determinant of the set of simultaneous linear equations is zero. By simple calculations we obtain for the determinant

$$2e^{-2\kappa a} \left(\alpha \sin(\alpha a) - \kappa \cos(\alpha a) \right) \left(\alpha \cos(\alpha a) + \kappa \sin(\alpha a) \right). \quad (33.5)$$

This can be zero only if one of the two expressions in brackets is zero. This implies that one of the following equations has to hold:

$$\kappa = \alpha \tan(\alpha a), \qquad\qquad (33.6)$$

$$\kappa = -\alpha \cot(\alpha a).$$

From these transcendental equations the energy eigenvalues E can be determined by numerical or graphical methods. If the first equation of (33.6) is satisfied, then the solution of (33.4) is

$$b_2 = 0 \quad \text{and} \quad b_1 = c_3 = c_2 e^{\kappa a} \cos(\alpha a), \qquad (33.7)$$

whereas if the second equation of (33.6) is satisfied:

$$c_2 = 0 \quad \text{and} \quad -b_1 = c_3 = b_2 e^{\kappa a} \sin(\alpha a). \qquad (33.8)$$

96

In the first case the eigenfunction of $\psi(x)$ is

$$\psi(x) = b_1 e^{\kappa x}, \qquad\qquad \text{if} \quad x < -a,$$

$$\psi(x) = b_1 e^{-\kappa a} \frac{\cos(\alpha x)}{\cos(\alpha a)}, \qquad \text{if} \quad -a < x < a, \qquad (33.9)$$

$$\psi(x) = b_1 e^{-\kappa x}, \qquad\qquad \text{if} \quad x > a.$$

This eigenfunction is an even function of x: $\psi(-x) = \psi(x)$. In the second case the eigenfunction is

$$\psi(x) = b_1 e^{\kappa x}, \qquad\qquad \text{if} \quad x < -a,$$

$$\psi(x) = -b_1 e^{-\kappa a} \frac{\sin(\alpha x)}{\sin(\alpha a)}, \qquad \text{if} \quad -a < x < a, \qquad (33.10)$$

$$\psi(x) = -b_1 e^{-\kappa x}, \qquad\qquad \text{if} \quad x > a.$$

The eigenfunction is now an odd function of x: $\psi(-x) = -\psi(x)$.
The amplitude b_1 of $\psi(x)$ can be calculated from the normalization condition:

$$\int_{-\infty}^{+\infty} |\psi(x)|^2 \, dx = 1. \qquad (33.11)$$

For both (33.9) and (33.10) we find that

$$\frac{1}{b_1^2} = ae^{-2\kappa a}\left(1 + \frac{1}{\kappa a}\right)\left(1 + \frac{\kappa^2}{\alpha^2}\right). \qquad (33.12)$$

Finally a very simple graphical method is introduced for evaluating the transcendental equation (33.6). According to equations (33.2) defining κ and α we have

$$\kappa^2 + \alpha^2 = \frac{8\pi^2 m_0}{h^2} V_0. \qquad (33.13)$$

Figure 13 shows the curves corresponding to the equations of (33.6) in a coordinate system of $\xi = \kappa a$, $\eta = \alpha a$. The solution of the eigenvalue problem is obtained by determining the points of intersection of these curves with the arc plotted according to (33.13). (Only the points of intersection falling into the first quadrant of the plane are of interest, as $\kappa > 0$ and $\alpha > 0$.) The energy eigenvalues can be determined from the ξ

coordinates of the points of intersection by the equation

$$E = - \frac{h^2 \zeta^2}{8\pi^2 m_0 a^2} .$$

It can be seen from the construction of Fig. 13 that the number of points of intersection (and thus of the energy eigenvalues) is always finite, and the larger the constant

$$\frac{8\pi^2 m_0 a^2 V_0}{h^2}$$

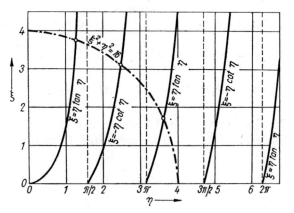

FIG. 13. A graphical solution of eqs. (33.6)

characteristic of the potential well, the greater the number of intersections. If

$$\frac{n\pi}{2} < \sqrt{\frac{8\pi^2 m_0 a^2 V_0}{h^2}} < \frac{(n+1)\pi}{2} , \tag{33.14}$$

then the number of intersections is $n + 1$. At least one even eigensolution can always be found, an odd eigenfunction only if

$$\sqrt{\frac{8\pi^2 m_0 a^2 V_0}{h^2}} > \frac{\pi}{2} .$$

The eigenfunction corresponding to the ground state is always even, in the excited states the even and odd eigenfunctions keep alternating.

98

34. Transmission of Particles across a Square Potential Well

In the preceding section the bound states developing in a square potential well were discussed. These states have negative energy values. The present section deals with the penetration of particles through the potential well. Let us assume that the particles come from the left-hand side, from the direction of the negative x-axis. Their energy should be a given positive value E determined by the initial velocity. The question is what the influence of the potential well on the particles is: what part of them will penetrate and what part will be reflected? Our starting point is again the Schrödinger equation (33.1). The general solution for positive E values is

$$\psi(x) = b_1 e^{ikx} + c_1 e^{-ikx}, \qquad \text{if } x < -a,$$

$$\psi(x) = b_2 e^{i\alpha x} + c_2 e^{-i\alpha x}, \qquad \text{if } -a < x < a, \qquad (34.1)$$

$$\psi(x) = b_3 e^{ikx} + c_3 e^{-ikx}, \qquad \text{if } x > a,$$

where α is the same as in (33.2) and

$$k = \sqrt{\frac{8\pi^2 m_0}{h^2} E}. \qquad (34.2)$$

The waves with amplitudes b_1, b_2 and b_3 travel towards the right in the direction of the positive x-axis, those with amplitudes c_1, c_2 and c_3 towards the left. As the incident particles come to the potential well from the left, in the region of $x > a$ only particles moving towards the right can be present, and thus $c_3 = 0$. The further amplitudes are again determined by the condition that $\psi(x)$ and $d\psi/dx$ have to be continuous at $x = -a$ and $x = a$. These continuity conditions, written out in detail, yield the following relations:

$$b_1 e^{-ika} + c_1 e^{ika} = b_2 e^{-i\alpha a} + c_2 e^{i\alpha a},$$

$$k(b_1 e^{-ika} - c_1 e^{ika}) = \alpha(b_2 e^{-i\alpha a} - c_2 e^{i\alpha a}), \qquad (34.3)$$

$$b_3 e^{ika} = b_2 e^{i\alpha a} + c_2 e^{-i\alpha a},$$

$$kb_3 e^{ika} = \alpha(b_2 e^{i\alpha a} - c_2 e^{-i\alpha a}).$$

Eliminating b_2 and c_2 from these equations, the following relations are obtained for the amplitudes of the incident (b_1), reflected (c_1) and transmitted wave (b_3):

$$b_1 e^{-ika} + c_1 e^{ika} = \frac{1}{2} e^{ika} \left[\left(1 + \frac{k}{\alpha} \right) e^{-2i\alpha a} + \left(1 - \frac{k}{\alpha} \right) e^{2i\alpha a} \right] b_3 ,$$

(34.4)

$$b_1 e^{-ika} - c_1 e^{ika} = \frac{1}{2} e^{ika} \left[\left(1 + \frac{\alpha}{k} \right) e^{-2i\alpha a} + \left(1 - \frac{\alpha}{k} \right) e^{2i\alpha a} \right] b_3 .$$

The transmission coefficient of the potential well is the probability that the particle can penetrate it. This probability is determined by the relation of the square of the absolute value of the amplitude of the incident and transmitted waves.

Thus the transmission coefficient is

$$T = \left| \frac{b_3}{b_1} \right|^2 .$$

From eq. (34.4) the following formula is obtained for T:

$$T = \frac{1}{\cos^2 (2\alpha a) + \frac{1}{4} \left(\frac{k}{\alpha} + \frac{\alpha}{k} \right)^2 \sin^2 (2\alpha a)} .$$

(34.5)

The probability of reflection from the potential well is

$$R = \left| \frac{c_1}{b_1} \right|^2 = \frac{\left[\frac{1}{4} \left(\frac{k}{\alpha} + \frac{\alpha}{k} \right)^2 - 1 \right] \sin^2 (2\alpha a)}{\cos^2 (2\alpha a) + \frac{1}{4} \left(\frac{k}{\alpha} + \frac{\alpha}{k} \right)^2 \sin^2 (2\alpha a)} .$$

(34.6)

(These two quantities are not independent of each other: the relation $T + R = 1$ holds.) Figures 14 and 15 show the transmission coefficient T as function of the energy E of incident particles for the case of two different potential wells. It can be seen that the maximum transmission of 100% is realized only in case of some discrete energy values, when $\sin (2\alpha a) = 0$, thus if $\alpha = n\pi/2a$. As displayed in Fig. 15, if the potential well is very deep and broad, that is if

$$\frac{8\pi^2 m_0 a^2 V_0}{h^2} \gg 1 ,$$

then the maxima of the transmission coefficient are very sharp. In these cases the maxima are called resonances.

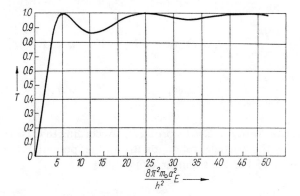

FIG. 14. The transmission coefficient of the square potential well, as a function of E, for the case where

$$\frac{8\pi^2 m_0 a^2 V_0}{h^2} = 16$$

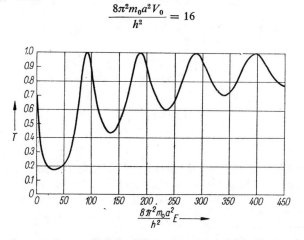

FIG. 15. The transmission coefficient of the square potential well, as a function of E, for the case where

$$\frac{8\pi^2 m_0 a^2 V_0}{h^2} = 800$$

35. Penetration of a Square Potential Barrier. The "Tunnel" Effect

In this section we consider the penetration of a mass point m_0 through the potential barrier shown in Fig. 16. The motion is also in this case one-dimensional; the Schrödinger equation of the problem is the following:

$$-\frac{h^2}{8\pi^2 m_0}\frac{d^2\psi}{dx^2} = E\psi, \qquad \text{if } x < -a,$$

$$-\frac{h^2}{8\pi^2 m_0}\frac{d^2\psi}{dx^2} + V_0\psi = E\psi, \quad \text{if } -a < x < a, \qquad (35.1)$$

$$-\frac{h^2}{8\pi^2 m_0}\frac{d^2\psi}{dx^2} = E\psi, \qquad \text{if } x > a.$$

Let us first deal with the $E < V_0$ problem. (According to classical mechanics in this case the particle rebounds from the potential barrier.) The general solution of the Schrödinger equation is

$$\psi(x) = b_1 e^{ikx} + c_1 e^{-ikx}, \qquad \text{if } x < -a,$$

$$\psi(x) = b_2 e^{\beta x} + c_2 e^{-\beta x}, \qquad \text{if } -a < x < a, \qquad (35.2)$$

$$\psi(x) = b_3 e^{ikx} + c_3 e^{-ikx}, \qquad \text{if } x > a,$$

where

$$k = \sqrt{\frac{8\pi^2 m_0}{h^2}E}$$

and

$$\beta = \sqrt{\frac{8\pi^2 m_0}{h^2}(V_0 - E)}.$$

If the particle comes from the left-hand side, in the region $x > a$ only a matter wave travelling towards the right can be present, thus $c_3 = 0$. From the continuity conditions for $x = -a$ and $x = a$ we find

$$b_1 e^{-ika} + c_1 e^{ika} = b_2 e^{-\beta a} + c_2 e^{\beta a},$$

$$ik(b_1 e^{-ika} - c_1 e^{ika}) = \beta(b_2 e^{-\beta a} - c_2 e^{\beta a}),$$
$$(35.3)$$

$$b_3 e^{ika} = b_2 e^{\beta a} + c_2 e^{-\beta a},$$

$$ikb_3 e^{ika} = \beta(b_2 e^{\beta a} - c_2 e^{-\beta a}).$$

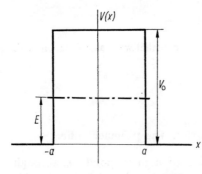

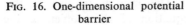

FIG. 16. One-dimensional potential barrier

The transmission coefficient of the potential barrier can be determined from the amplitude ratio of the incident and transmitted wave. We find from eq. (35.3) by a simple calculation that

$$T = \left| \frac{b_3}{b_1} \right|^2 = \frac{1}{\cosh^2(2\beta a) + \dfrac{1}{4} \left(\dfrac{\beta}{k} - \dfrac{k}{\beta} \right)^2 \sinh^2(2\beta a)}. \quad (35.4)$$

This result is highly remarkable. The particles may penetrate the potential barrier with a finite, non-zero probability, even if classical mechanics predicts a reflection. This effect is known as the *tunnel effect*. The probability of the tunnel effect decreases rapidly as the width of the barrier increases; if $\beta a \gg 1$ the following approximate expression for the transmission coefficient T holds:

$$T = \frac{16E(V_0 - E)}{V_0^2} e^{-4\beta a}. \quad (35.5)$$

Let us deal briefly with the case when $E > V_0$. According to classical mechanics in this case the particle always passes across the barrier. Calculations similar to the previous ones yield for the transmission coefficient:

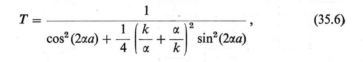

$$T = \frac{1}{\cos^2(2\alpha a) + \dfrac{1}{4} \left(\dfrac{k}{\alpha} + \dfrac{\alpha}{k} \right)^2 \sin^2(2\alpha a)}, \quad (35.6)$$

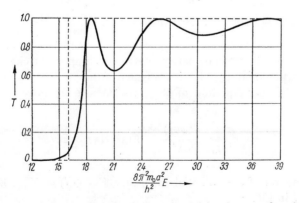

FIG. 17. The transmission coefficient of the potential barrier as a function of E, for the case where

$$\frac{8\pi^2 m_0 a^2 V_0}{h^2} = 16$$

where

$$k = \sqrt{\frac{8\pi^2 m_0}{h^2} E} \quad \text{and} \quad \alpha = \sqrt{\frac{8\pi^2 m_0}{h^2} (E - V_0)}. \quad (35.7)$$

That shows that T reaches the classical mechanical value of $T = 1$ only at some resonance energies which are determined by the condition $\sin(2\alpha a) = 0$, thus $\alpha = n\pi/2a$. Figure 17 illustrates the transmission coefficient T, calculated from eqs. (35.4) and (35.6), as a function of the energy of the incident particles for the case of a potential barrier, where

$$\frac{8\pi^2 m_0 a^2 V_0}{h^2} = 16 .$$

36. Periodic Potential

The conduction electrons of a metal crystal move in the potential field produced by the metal ions. The most striking characteristic of this potential is that it varies periodically with the lattice constant. Figure 18 shows a one-dimensional periodic potential consisting of a regular repetition of square potential wells. Its period is $l = 2a_1 + 2a_2$. This potential is a very idealized model of the potential distribution within a metal; nevertheless every important characteristic of a periodic potential can be investigated from it.

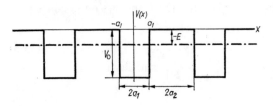

FIG. 18. One-dimensional periodic potential

The Schrödinger equation of the one-dimensional movement is

$$-\frac{h^2}{8\pi^2 m_0} \frac{d^2\psi(x)}{dx^2} + V(x)\,\psi(x) = E\psi(x) . \quad (36.1)$$

If l is the period of the potential $V(x)$, then $V(x + l) = V(x)$. By replacing x by $x + l$ in (36.1)

$$-\frac{h^2}{8\pi^2 m_0} \frac{d^2\psi(x + l)}{dx^2} + V(x)\,\psi(x + l) = E\psi(x + l) . \quad (36.2)$$

Thus if $\psi(x)$ is a solution of the Schrödinger equation, $\psi(x + l)$ is a solution as well. On the other hand, in the case of a one-dimensional motion only a single eigenfunction can belong to a given E eigenvalue, and thus $\psi(x)$ and $\psi(x + l)$ can differ from each other only by a constant factor.

$$\psi(x + l) = C\psi(x) . \tag{36.3}$$

If this rule is applied n times we find as a result $\psi(x + nl) = C^n\psi(x)$. As the wave function ψ has to remain finite along the whole x-axis C^n cannot become infinite even when $n \to \pm\infty$. This is only possible if C is a phase factor of absolute value unity:

$$C = e^{i\Theta} . \tag{36.4}$$

Here Θ is a real number which can be supposed to satisfy, without any restriction

$$-\pi < \Theta < \pi . \tag{36.5}$$

Let us now deal with the solution of the Schrödinger equation for the case of the special periodic potential shown in Fig. 18.

According to rule (36.3) it is enough to investigate a section of length l. This could be the section $-a_1 < x < a_1 + 2a_2$. The Schrödinger equation is

$$-\frac{h^2}{8\pi^2 m_0} \frac{d^2\psi}{dx^2} - V_0\psi = E\psi , \quad \text{if } -a_1 < x < a_1$$

$$-\frac{h^2}{8\pi^2 m_0} \frac{d^2\psi}{dx^2} = E\psi , \qquad \text{if } a_1 < x < a_1 + 2a_2 . \tag{36.6}$$

First let us suppose that $-V_0 < E < 0$. Introducing the notation

$$\alpha = \sqrt{\frac{8\pi^2 m_0}{h^2}(E + V_0)} \quad \text{and} \quad \kappa = \sqrt{-\frac{8\pi^2 m_0}{h^2}E} \tag{36.7}$$

the general solution is

$$\psi(x) = b_1 e^{i\alpha x} + c_1 e^{-i\alpha x}, \quad \text{if } -a_1 < x < a_1 ,$$

$$\psi(x) = b_2 e^{\kappa x} + c_2 e^{-\kappa x}, \quad \text{if } a_1 < x < a_1 + 2a_2 . \tag{36.8}$$

It is known that ψ and $d\psi/dx$ must vary continuously also on those places where the potential shows a discontinuity. The condition of continuity at

105

$x = a_1$ is

$$b_1 e^{i\alpha a_1} + c_1 e^{-i\alpha a_1} = b_2 e^{\kappa a_1} + c_2 e^{-\kappa a_1},$$

$$i\alpha(b_1 e^{i\alpha a_1} - c_1 e^{-i\alpha a_1}) = \kappa(b_2 e^{\kappa a_1} - c_2 e^{-\kappa a_1}).$$

(36.9)

The potential shows a second discontinuity at $x = a_1 + 2a_2$. To the right of this place the eigenfunction can be determined from the first equation of (36.8) by the help of rules (36.3) and (36.4)

$$\psi(x) = e^{i\Theta}(b_1 e^{i\alpha(x-l)} + c_1 e^{-i\alpha(x-l)}),$$

(36.10)

$$a_1 + 2a_2 < x < 3a_1 + 2a_2.$$

This has to be joined continuously with a continuous first differential quotient to the function defined by the second equation of (36.8). From this the following relations follow:

$$b_2 e^{\kappa(a_1+2a_2)} + c_2 e^{-\kappa(a_1+2a_2)} = e^{i\Theta}(b_1 e^{-i\alpha a_1} + c_1 e^{i\alpha a_1}),$$

$$\kappa(b_2 e^{\kappa(a_1+2a_2)} - c_2 e^{-\kappa(a_1+2a_2)}) = i\alpha e^{i\Theta}(b_1 e^{-i\alpha a_1} - c_1 e^{i\alpha a_1}).$$

(36.11)

Equations (36.9) and (36.11) form a set of simultaneous homogeneous linear equations from which the constants b_1, c_1, b_2 and c_2 can be determined. A non-trivial solution exists only if the determinant of the set of equations is equal to zero:

$$\begin{vmatrix} e^{i\alpha a} & e^{-i\alpha a_1} & -e^{\kappa a_1} & -e^{-\kappa a_1} \\ i\alpha e^{i\alpha a_1} & -i\alpha e^{-i\alpha a_1} & -\kappa e^{\kappa a_1} & \kappa e^{-\kappa a_1} \\ e^{i(\Theta-\alpha a_1)} & e^{i(\Theta+\alpha a_1)} & -e^{\kappa(a_1+2a_2)} & -e^{-\kappa(a_1+2a_2)} \\ i\alpha e^{i(\Theta-\alpha a_1)} & -i\alpha e^{i(\Theta+\alpha a_1)} & -\kappa e^{\kappa(a_1+2a_2)} & \kappa e^{-\kappa(a_1+2a_2)} \end{vmatrix} = 0.$$

Evaluating the determinant we find

$$\cosh(2\kappa a_2)\cos(2\alpha a_1) + \frac{\kappa^2 - \alpha^2}{2\kappa\alpha}\sinh(2\kappa a_2)\sin(2\alpha a_1) = \cos\Theta. \quad (36.12)$$

On the right-hand side stands the cosine of a real number Θ, on the left-hand side, if the definitions (36.7) of α and κ are taken into consideration, a transcendent function of energy E is found. Let us call this function $f(E)$. Then the following relation holds:

$$f(E) = \cos\Theta. \quad (36.13)$$

This can be true only if

$$-1 \le f(E) \le 1 . \tag{36.14}$$

The eigenvalues of the Schrödinger equation are those E energies for which this inequality holds. The function $f(E)$ is shown in Fig. 19 for

$$a_1 = 2a_2 \quad \text{and} \quad \frac{8\pi^2 m_0 a^2 V_0}{h^2} = 16 .$$

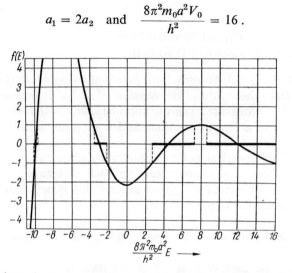

FIG. 19. The determination of the energy bands for the case of a periodic potential

We see that the eigenvalues E satisfying the condition (36.14) form intervals, so-called bands or zones, of finite length, separated by forbidden zones. (It is easy to prove that, for example, those E energies for which $2\alpha a_1/\pi$ is an integer are all forbidden, as in this case $|f(E)| = \cosh{(2\kappa a_2)}$, so that the inequality (36.14) cannot be fulfilled. It follows from the continuity of $f(E)$ that the forbidden zones form the surroundings of these energy values.)

The case of $E > 0$ can be dealt with quite similarly. Then instead of (36.12) the following equation serves for the determination of the energy bands:

$$f(E) = \cos{(2ka_2)}\cos{(2\alpha a_1)} - \frac{k^2 + \alpha^2}{2k\alpha}\sin{(2ka_2)}\sin{(2\alpha a_1)} = \cos{\Theta} \tag{36.15}$$

where

$$k = \sqrt{\frac{8\pi^2 m_0}{h^2} E} . \tag{36.16}$$

107

In Fig. 19 the energy bands falling on the positive E-axis are defined according to (36.15).

The band structure of the energy eigenvalues is one of the most important characteristics of the periodic crystal lattices. The fundamental properties of solids can be interpreted on this basis, and this has become the start-point of the quantum theory of electric conduction.

37. Movement in a Central Potential Field (Spherically Symmetrical Potential)

A potential field is called central if the potential V depends only on the radial distance r from the centre of the potential field but is independent of the direction of this distance. The quantum-mechanical treatment of a mass point moving in a central potential field is of great practical importance both in atomic and nuclear physics.

To introduce spherical polar coordinates and express the Schrödinger equation in them is often convenient for such problems.

The transformation equations connecting this system with rectangular Cartesian coordinates are:

$$x = r \sin \vartheta \cos \phi, \quad y = r \sin \vartheta \sin \phi, \quad z = r \cos \vartheta. \quad (37.1)$$

From this the Laplace operator is obtained as

$$\nabla^2 = \frac{\partial^2}{\partial x^2} + \frac{\partial^2}{\partial y^2} + \frac{\partial^2}{\partial z^2} = \frac{\partial^2}{\partial r^2} + \frac{2}{r} \frac{\partial}{\partial r}$$

$$+ \frac{1}{r^2} \left(\frac{\partial^2}{\partial \vartheta^2} + \cot \vartheta \frac{\partial}{\partial \vartheta} + \frac{1}{\sin^2 \vartheta} \frac{\partial^2}{\partial \phi^2} \right). \quad (37.2)$$

Naturally the centre of the potential field is chosen as the origin of the polar coordinate system, in this case $V = V(r)$. Thus the Schrödinger equation in the polar coordinates is

$$\frac{\partial^2 \psi}{\partial r^2} + \frac{2}{r} \frac{\partial \psi}{\partial r} + \frac{1}{r^2} \left(\frac{\partial^2 \psi}{\partial \vartheta^2} + \cot \vartheta \frac{\partial \psi}{\partial \vartheta} + \frac{1}{\sin^2 \vartheta} \frac{\partial^2 \psi}{\partial \phi^2} \right)$$

$$+ \frac{8\pi^2 m_0}{h^2} (E - V(r)) \psi = 0. \quad (37.3)$$

We assume that

$$\psi = R(r) Y(\vartheta, \phi), \quad (37.4)$$

where R is a function of r only and Y is a function of the polar angles ϑ and ϕ. Substituting this expression into the Schrödinger equation and multiplying both sides of the equation by r^2/ψ, we obtain

$$\frac{r^2}{R} \left(\frac{d^2R}{dr^2} + \frac{2}{r} \frac{dR}{dr} \right) + \frac{8\pi^2 m_0}{h^2} r^2(E - V(r))$$

$$= -\frac{1}{Y} \left(\frac{\partial^2 Y}{\partial \vartheta^2} + \cot \vartheta \frac{\partial Y}{\partial \vartheta} + \frac{1}{\sin^2 \vartheta} \frac{\partial^2 Y}{\partial \phi^2} \right). \quad (37.5)$$

The expression on the left-hand side of the equation depends only on r, that on the right-hand side only on the polar angles ϑ and ϕ. The equation can hold for every arbitrary value of r, ϑ and ϕ only if both sides are independently constant. Let us call this constant λ. Then the following two equations are obtained

$$\frac{d^2R}{dr^2} + \frac{2}{r} \frac{dR}{dr} - \frac{\lambda}{r^2} R + \frac{8\pi^2 m_0}{h^2} (E - V(r)) R = 0, \quad (37.6)$$

$$\frac{\partial^2 Y}{\partial \vartheta^2} + \cot \vartheta \frac{\partial Y}{\partial \vartheta} + \frac{1}{\sin^2 \vartheta} \frac{\partial^2 Y}{\partial \phi^2} + \lambda Y = 0. \quad (37.7)$$

The first equation is a normal differential equation for the function $R(r)$, it is called the *radial Schrödinger equation*. To solve this equation the concrete form of the $V(r)$ potential function has to be given.

On the other hand, eq. (37.7), concerning the function $Y(\vartheta, \phi)$, does not imply the potential, and so it gives the angle-dependent part of ψ for every central potential field.

Let us assume that

$$Y(\vartheta, \phi) = T(\vartheta) F(\phi). \quad (37.8)$$

Substituting this into (37.7), by multiplying with $\sin^2\vartheta/Y$ the expressions depending on ϑ and ϕ can be separated:

$$\frac{\sin^2 \vartheta}{T} \left(\frac{d^2T}{d\vartheta^2} + \cot \vartheta \frac{dT}{d\vartheta} \right) + \lambda \sin^2 \vartheta = -\frac{1}{F} \frac{d^2F}{d\phi^2}. \quad (37.9)$$

Here it can be stated again that both sides of the equation have to be independently constant, since otherwise, the equality cannot hold for any arbitrary value of the polar angles. And so the following two equations are found:

$$\frac{d^2T}{d\vartheta^2} + \cot\vartheta\,\frac{dT}{d\vartheta} + \left(\lambda - \frac{\mu}{\sin^2\vartheta}\right) T = 0\,, \tag{37.10}$$

$$\frac{d^2F}{d\phi^2} + \mu F = 0\,. \tag{37.11}$$

The solution of the equation for $F(\phi)$ is

$$F(\phi) = e^{i\sqrt{\mu}\,\phi}\,.$$

As we require that ψ should be an eigenfunction, and thus $F(\phi)$ should be a single-valued function of space, F has to be periodic in ϕ with a period of 2π, and this implies that $m = \sqrt{\mu}$ has to be an integer. Thus,

$$Y(\vartheta, \phi) = T(\vartheta)\,e^{im\phi}, \quad m = 0, \pm 1, \pm 2, \ldots \tag{37.12}$$

Replacing μ by m^2 in eq. (37.10) serving for the definition of function $T(\vartheta)$ and performing the substitution $x = \cos\vartheta$, by a brief calculation we obtain

$$(1 - x^2)\frac{d^2P}{dx^2} - 2x\frac{dP}{dx} + \left(\lambda - \frac{m^2}{1 - x^2}\right) P = 0\,, \tag{37.13}$$

where $P(\cos\vartheta)$ is written instead of $T(\vartheta)$.

This equation is the so-called *spherical Bessel differential equation*. From the theory of the spherical equations it follows that (37.13) has a regular solution at $x = \pm 1$ (thus at $\vartheta = 0$ and $\vartheta = \pi$) only if λ is of the following form:

$$\lambda = (k + |m|)\,(k + |m| + 1)\,,$$

where k is a non-negative integer ($k = 0, 1, 2, \ldots$). Denoting the integer $k + |m|$ by l, then $\lambda = l(l + 1)$. The solutions of the equation for $P(x)$, belonging to $\lambda = l(l + 1)$ are called *spherical functions of the degree l*. Their number is $2l + 1$, because for a given l, m can take the values 0, $\pm 1, \pm 2, \ldots, \pm l$. If $m = 0$, the so-called normal spherical functions of the degree l are received. These are simple polynomials of the degree l, and are denoted by $P_l(x)$. It can be proved that the polynomials p_l are equal to the coefficients of the power series in r of the function $G(r, x) = (1 - 2rx + r^2)^{-\frac{1}{2}}$, thus

$$(1 - 2rx + r^2)^{-\frac{1}{2}} = \sum_{l=0}^{\infty} P_l(x)\,r^l.$$

From this we find by simple calculation that

$$P_0(x) = 1, \qquad\qquad P_3(x) = \frac{1}{2}(5x^3 - 3x),$$

$$P_1(x) = x, \qquad\qquad P_4(x) = \frac{1}{8}(35x^4 - 30x^2 + 3),$$

$$P_2(x) = \frac{1}{2}(3x^2 - 1), \qquad P_5(x) = \frac{1}{8}(63x^5 - 70x^3 + 15x).$$

All this concerned the case when $m = 0$. The solutions of eq. (37.13) can be expressed in terms of the polynomials $P_l(x)$ also if $m \neq 0$. We can see by direct substitution that when $m \neq 0$

$$P_l^m(x) = (1 - x^2)^{|m|/2} \frac{d^{|m|}P_l(x)}{dx^{|m|}} \tag{37.14}$$

satisfies the differential equation. The $P_l^m(x)$ functions are called adjoint spherical functions.

As a final result the solutions of eq. (37.7) are

$$Y_l^m(\vartheta, \phi) = A_{lm} P_l^m(\cos \vartheta)\, e^{im\phi}, \tag{37.15}$$

where l is a non-negative integer, m is an integer, too, for which $|m| \leq l$. The constant A_{lm} is a normalizing coefficient chosen such that

$$\int_0^\pi \int_0^{2\pi} |Y_l^m(\vartheta, \phi)|^2 \sin \vartheta\, d\vartheta\, d\phi = 1. \tag{37.16}$$

The expressions of the first few normalized spherical functions $Y_l^m(\vartheta, \phi)$ are readily found to be

$$Y_0^0 = \left(\frac{1}{4\pi}\right)^{\frac{1}{2}},$$

$$Y_1^0 = \left(\frac{3}{4\pi}\right)^{\frac{1}{2}} \cos \vartheta, \qquad\qquad Y_1^{\pm 1} = \left(\frac{3}{8\pi}\right)^{\frac{1}{2}} \sin \vartheta\, e^{\pm i\phi},$$

$$Y_2^0 = \left(\frac{5}{16\pi}\right)^{\frac{1}{2}} (3\cos^2 \vartheta - 1), \qquad Y_3^0 = \left(\frac{7}{16\pi}\right)^{\frac{1}{2}} (5\cos^3 \vartheta - 3\cos \vartheta),$$

$$Y_2^{\pm 1} = \left(\frac{15}{8\pi}\right)^{\frac{1}{2}} \cos \vartheta \sin \vartheta\, e^{\pm i\phi}, \qquad Y_3^{\pm 1} = \left(\frac{21}{64\pi}\right)^{\frac{1}{2}} (5\cos^2 \vartheta - 1) \sin \vartheta\, e^{\pm i\phi},$$

$$Y_2^{\pm 2} = \left(\frac{15}{32\pi}\right)^{\frac{1}{2}} \sin^2 \vartheta\, e^{\pm 2i\phi}, \qquad Y_3^{\pm 2} = \left(\frac{105}{32\pi}\right)^{\frac{1}{2}} \cos \vartheta \sin^2 \vartheta\, e^{\pm 2i\phi},$$

$$Y_3^{\pm 3} = \left(\frac{35}{64\pi}\right)^{\frac{1}{2}} \sin^3 \vartheta\, e^{\pm 3i\phi}.$$

Because $\lambda = l(l + 1)$ the radial Schrödinger equation (37.6) for the orbital quantum number l will take the following form:

$$\frac{d^2 R}{dr^2} + \frac{2}{r} \frac{dR}{dr} - \frac{l(l+1)}{r^2} R + \frac{8\pi^2 m_0}{h^2} (E - V(r)) R = 0 \, . \quad (37.17)$$

38. The Moment of Momentum (or Angular Momentum)

Both in the classical and quantum-mechanical investigations of the three-dimensional movement of a mass point the concept of orbital moment of momentum is of great importance. The moment of momentum belonging to the orbital movement has already been dealt with in Section 27. It has been shown that the operators assigned to the Cartesian coordinates of the angular momentum are the following:

$$N_x = \frac{h}{2\pi i} \left(y \frac{\partial}{\partial z} - z \frac{\partial}{\partial y} \right),$$

$$N_y = \frac{h}{2\pi i} \left(z \frac{\partial}{\partial x} - x \frac{\partial}{\partial z} \right),$$

$$N_z = \frac{h}{2\pi i} \left(x \frac{\partial}{\partial y} - y \frac{\partial}{\partial x} \right).$$

Let us introduce the spherical polar coordinates by the transformations (36.1). The transformation rule of the differential operators in this case is

$$\frac{\partial}{\partial x} = \sin \vartheta \cos \phi \frac{\partial}{\partial r} + \frac{\cos \vartheta \cos \phi}{r} \frac{\partial}{\partial \vartheta} - \frac{\sin \phi}{r \sin \vartheta} \frac{\partial}{\partial \phi},$$

$$\frac{\partial}{\partial y} = \sin \vartheta \sin \phi \frac{\partial}{\partial r} + \frac{\cos \vartheta \sin \phi}{r} \frac{\partial}{\partial \vartheta} + \frac{\cos \phi}{r \sin \vartheta} \frac{\partial}{\partial \phi},$$

$$\frac{\partial}{\partial z} = \cos \vartheta \frac{\partial}{\partial r} - \frac{\sin \vartheta}{r} \frac{\partial}{\partial \vartheta}.$$

For the angular momentum operators we find the expressions

$$N_x = \frac{h}{2\pi i} \left(- \sin \phi \frac{\partial}{\partial \vartheta} - \cot \vartheta \cos \phi \frac{\partial}{\partial \phi} \right),$$

$$N_y = \frac{h}{2\pi i} \left(\cos \phi \frac{\partial}{\partial \vartheta} - \cot \vartheta \sin \phi \frac{\partial}{\partial \phi} \right), \quad (38.1)$$

$$N_z = \frac{h}{2\pi i} \frac{\partial}{\partial \phi}$$

and the square of the angular momentum is

$$N^2 = - \frac{h^2}{4\pi^2} \left(\frac{\partial^2}{\partial \vartheta^2} + \cot \vartheta \frac{\partial}{\partial \vartheta} + \frac{1}{\sin^2 \vartheta} \frac{\partial^2}{\partial \phi^2} \right). \tag{38.2}$$

It is obvious that the spherical functions Y_l^m are the eigenfunctions of the N_z and N^2 operators. As Y_l^m contains the azimuthal angle ϕ in the form $e^{im\phi}$, the effect of the N_z operator is

$$N_z Y_l^m(\vartheta, \phi) = \frac{h}{2\pi i} \frac{\partial Y_l^m}{\partial \phi} = \frac{mh}{2\pi} Y_l^m(\vartheta, \phi). \tag{38.3}$$

Thus the result is that the N_z operator changes Y_l^m into $mh/2\pi$ times Y_l^m. This implies that the eigenvalues of N_z are

$$\frac{mh}{2\pi}, \tag{38.4}$$

where

$$m = 0, \pm 1, \pm 2, \ldots, \pm l.$$

Thus the quantum number m gives the length of the component of the orbital angular momentum in the direction of the z-axis. As shown by the above-mentioned result this component is a (positive or negative) integer multiple of $h/2\pi$. m is equal to the magnetic quantum number introduced in Bohr's theory.

The action of the N^2 operator on Y_l^m can be determined by comparing formulae (38.2) and (37.7). Writing $\lambda = l(l + 1)$ in the latter, we obtain

$$N^2 Y_l^m(\vartheta, \phi) = \frac{l(l + 1) h^2}{4\pi^2} Y_l^m(\vartheta, \phi). \tag{38.5}$$

Thus the spherical functions Y_l^m are the eigenfunctions of the N^2 operator as well. The eigenvalues are

$$\frac{l(l + 1) h^2}{4\pi^2}, \tag{38.6}$$

where

$$l = 0, 1, 2, \ldots$$

From this it follows that the eigenvalues of N, the absolute values of the angular momentum are $\sqrt{l(l + 1)} \cdot h/2\pi$. The quantum number l determines the absolute value of the angular momentum, and it is equal to the orbital quantum number of the Bohr theory.

We mention that the eigenvalues (38.4) are valid not only for the components measured in the direction of the z-axis, but also for those in the x- and y-directions. (This is obvious, as the z-axis cannot be singled out from other directions of space.) The effect of the components N_x and N_y on the Y_l^m spherical functions can be calculated from formulae (38.1), (37.15) and (37.14). The result is the following:

$$N_x Y_l^m = \frac{h}{4\pi} \sqrt{(l-m)(l+m+1)}\, Y_l^{m+1} + \frac{h}{4\pi} \sqrt{(l+m)(l-m+1)}\, Y_l^{m-1},$$

(38.7)

$$N_y Y_l^m = -\frac{ih}{4\pi} \sqrt{(l-m)(l+m+1)}\, Y_l^{m+1} + \frac{ih}{4\pi} \sqrt{(l+m)(l-m+1)}\, Y_l^{m-1}.$$

It can be seen that the spherical functions Y_l^m are not eigenfunctions of the N_x and N_y components. Linear combinations of the Y_l^m-s corresponding to different magnetic quantum numbers will be the eigenfunctions of these components. That the eigenfunctions of N_z are much simpler than those of N_x and N_y is the result of the fact that the z-axis was chosen as the axis of the polar coordinate system.

Let us now investigate the total angular momentum of an electron, composed of the orbital angular momentum and the spin:

$$\mathbf{J} = \mathbf{N} + \mathbf{S}.$$

(38.8)

It will be shown that for the quantization of $\mathbf{J}$ similar rules are valid as for the case of the orbital angular momentum. Thus the eigenvalues of J^2 are, by analogy with (38.6):

$$\frac{j(j+1)h^2}{4\pi^2},$$

(38.9)

where, however, the j, the total angular momentum quantum numbers, in contrast to l, are not integers but half integers: $j = 1/2,\ 3/2,\ 5/2,\ \ldots$ The eigenvalues of component J_z are, similar to (38.4),

$$\frac{m_j h}{2\pi},$$

(38.10)

where

$$m_j = \pm\frac{1}{2},\ \pm\frac{3}{2},\ \ldots,\ \pm j.$$

114

To prove (38.9) let us first compute the square of the operator $\mathbf{J} = \mathbf{N} + \mathbf{S}$:

$$J^2 = (\mathbf{N} + \mathbf{S})^2 = N^2 + S^2 + 2\mathbf{NS}.$$

Taking the results (28.12) and (28.22) for the spin operator $\mathbf{S}$ into account, we can write

$$J^2 = N^2 + \frac{3h^2}{16\pi^2} + \frac{h}{2\pi}(N_x\sigma_x + N_y\sigma_y + N_z\sigma_z).\qquad(38.11)$$

We shall denote the eigenfunctions of the total angular momentum by F, we thus have

$$J^2 F = \frac{\beta h^2}{4\pi^2} F,\qquad(38.12)$$

if the eigenvalue is written as $\beta h^2/4\pi^2$. The eigenfunction F contains, naturally, besides the space coordinates also the spin variable. Its general form can be put following (28.4):

$$F = F_+(x, y, z)\,\chi_{+\frac{1}{2}} + F_-(x, y, z)\,\chi_{-\frac{1}{2}}.\qquad(38.13)$$

If this is substituted into the eigenvalue equation (38.12) and the expression (38.11) for J^2 is taken into consideration, the following equation is reached:

$$\left[\left(N^2 + \frac{3h^2}{16\pi^2} + \frac{h}{2\pi}N_z\right)F_+ + \frac{h}{2\pi}(N_x - iN_y)F_-\right]\chi_{+\frac{1}{2}}$$

$$+\left[\left(N^2 + \frac{3h^2}{16\pi^2} - \frac{h}{2\pi}N_z\right)F_- + \frac{h}{2\pi}(N_x + iN_y)F_+\right]\chi_{-\frac{1}{2}} = \frac{\beta h^2}{4\pi^2}F_+\chi_{+\frac{1}{2}} + F_-\chi_{-\frac{1}{2}}).$$

$$(38.14)$$

Here Table 1 (see p. 77) showing the action of the Pauli matrices on the basic spinors χ_{m_s} has been used.

This equation can be valid for both values of the spin variable only if the coefficients of the spinors $\chi_{+\frac{1}{2}}$ and $\chi_{-\frac{1}{2}}$ are separately equal to each other on the left- and right-hand side. This leads to the following two equations:

$$\left(N^2 + \frac{3h^2}{16\pi^2} + \frac{h}{2\pi}N_z\right)F_+ + \frac{h}{2\pi}(N_x - iN_y)F_- = \frac{\beta h^2}{4\pi^2}F_+,$$

$$(38.15)$$

$$\left(N^2 + \frac{3h^2}{16\pi^2} - \frac{h}{2\pi}N_z\right)F_- + \frac{h}{2\pi}(N_x + iN_y)F_+ = \frac{\beta h^2}{4\pi^2}F_-.$$

This is a system of partial differential equations for the functions F_+ and F_-. It has to be emphasized that (38.15) does not contain the spin variable any more, only the expressions of the space coordinates.

Let us try to find the solution in the following form:

$$F_+ = C_+ Y_l^m(\vartheta, \phi), \qquad F_- = C_- Y_l^{m+1}(\vartheta, \phi), \qquad (38.16)$$

where C_+ and C_- are at the present undefined coefficients. The action of the Cartesian components N_x, N_y and N_z as well as that of N^2 on the spherical function Y_l^m is already known, the results are given by formulae (38.3) (38.7) and (38.5). Substituting the expression (38.16) into (38.15) we obtain

$$\left[l(l+1) + \frac{3}{4} + m \right] C_+ + \sqrt{(l+m+1)(l-m)}\, C_- = \beta C_+ ,$$

$$(38.17)$$

$$\left[l(l+1) + \frac{3}{4} - m - 1 \right] C_- + \sqrt{(l+m+1)(l-m)}\, C_+ = \beta C_- .$$

This is a set of simultaneous homogeneous linear equations for the coefficients C_+ and C_-. From the condition of solvability, i.e. from the condition that the determinant of the set of equations is zero, a quadratic equation is obtained for the parameter β determining the eigenvalue:

$$\beta^2 - 2\left(l + \frac{1}{2}\right)^2 \beta + \left(l^2 + l + \frac{1}{4}\right)\left(l^2 + l - \frac{3}{4}\right) = 0 . \quad (38.18)$$

It is interesting that this equation does not contain the magnetic quantum number m, as this drops out when forming the determinant. The roots of this equation are

$$\beta = l^2 + 2l + \frac{3}{4} = \left(l + \frac{1}{2}\right)\left(l + \frac{3}{2}\right),$$

and

$$\beta = l^2 - \frac{1}{4} = \left(l - \frac{1}{2}\right)\left(l + \frac{1}{2}\right).$$

Both solutions can be written in the form

$$\beta = j(j+1) \qquad (38.19)$$

in the first case $j = l + \frac{1}{2}$, in the second one $j = l - \frac{1}{2}$. Thus our statement (38.9) has been fully proved.

The $j = l + \frac{1}{2}$ value corresponds to the vectorial addition of the l orbital momentum and the $\frac{1}{2}$ spin, whereas the $j = l - \frac{1}{2}$ value corresponds to their subtraction. Writing the eigenvalues into the equations (38.17) the ratio of the coefficients C_+ and C_- can be determined by a simple calculation. If the eigenfunction is normalized in such a way that $C_+^2 + C_-^2 = 1$, then performing the simple calculations we find that

$$C_+ = \sqrt{\frac{l+m+1}{2l+1}}, \qquad C_- = \sqrt{\frac{l-m}{2l+1}} \qquad \text{if } j = l + \frac{1}{2},$$

$$\text{(38.20)}$$

$$C_+ = \sqrt{\frac{l-m}{2l+1}}, \qquad C_- = -\sqrt{\frac{l+m+1}{2l+1}}, \qquad \text{if } j = l - \frac{1}{2}.$$

The eigenfunction F is formed according to (38.13) and (38.16):

$$F = \sqrt{\frac{l+m+1}{2l+1}}\, Y_l^m \chi_{+\frac{1}{2}} + \sqrt{\frac{l-m}{2l+1}}\, Y_l^{m+1}\chi_{-\frac{1}{2}}, \quad \text{if } j = l + \frac{1}{2},$$

$$\text{(38.21)}$$

$$F = \sqrt{\frac{l-m}{2l+1}}\, Y_l^m \chi_{+\frac{1}{2}} - \sqrt{\frac{l+m+1}{2l+1}}\, Y_l^{m+1}\chi_{-\frac{1}{2}}, \quad \text{if } j = l - \frac{1}{2}.$$

Let us now determine the eigenvalues of the component J_z. The action of the operator $J_z = N_z + S_z$ onto the eigenfunctions (38.21) can be put down immediately on the basis of the rules

$$N_z Y_l^m = \frac{mh}{2\pi}\, Y_l^m \quad \text{and} \quad S_z \chi_{m_s} = \frac{m_s h}{2\pi}\, \chi_{m_s}.$$

The result for both j values is

$$J_z F = \left(m + \frac{1}{2}\right)\frac{h}{2\pi}\, F. \qquad \text{(38.22)}$$

The eigenvalue satisfies (38.10) if the $m_j = m + \frac{1}{2}$ abbreviation is introduced. If we also take into consideration that in the first function of (38.21) the m quantum number can take the values $m = -(l+1), -l, -(l-1), \ldots, -1, 0, 1, \ldots, l$, and in the second solution the values $m = -l, -(l-1), \ldots, -1, 0, 1, \ldots (l-1)$, then it can be seen that for both solutions m_j takes the values $-j, \ldots, -\frac{1}{2}, \frac{1}{2}, \ldots j$ of (38.10).

Summarizing these results, the functions F of (38.21) are the joint eigenfunctions of the N^2, J^2 and J_z operators. The corresponding eigenvalues are in turn $l(l + 1) h^2/4\pi^2$, $j(j + 1) h^2/4\pi^2$ and $m_j h/2\pi$. The possible values of the quantum numbers are

$$l = 0, 1, 2, 3, \ldots,$$

$$j = l + \frac{1}{2}, l - \frac{1}{2},$$

$$m_j = \pm \frac{1}{2}, \pm \frac{3}{2}, \ldots, \pm j. \tag{38.23}$$

These quantum numbers are written as indexes to the F eigenfunctions, so that the symbol of the eigenfunction of the angular momentum is F_{ljm_j}.

39. Magnetic Moment

According to the well-known law of electrodynamics the magnetic field of a charged particle with an electric charge of $-e$ moving along a closed orbit is equal to the field of a magnetic dipole with a dipole moment of

$$\mu_N = -\frac{e}{2m_0 c} \mathbf{N}. \tag{39.1}$$

In the case of an electron the magnetic moment μ_s connected with the spin has still to be added to this expression. This latter expression is, as shown earlier,

$$\mu_s = -\frac{e}{m_0 c} \mathbf{S}.$$

The total dipole moment of a moving electron thus is

$$\mu = \mu_N + \mu_s = -\frac{e}{2m_0 c} (\mathbf{N} + 2\mathbf{S}). \tag{39.2}$$

The eigenvalues and eigenfunctions of the z component of this magnetic moment can be easily calculated. Let us apply μ_z to the function $Y_l^m \chi_{m_s}$. We find that

$$\mu_z(Y_l^m \chi_{m_s}) = -\frac{e}{2m_0 c} (N_z + 2S_z)(Y_l^m \chi_{m_s}) = -\frac{eh}{4\pi m_0 c} (m + 2m_s)(Y_l^m \chi_{m_s}).$$

118

It can be seen that $Y_l^m \chi_{m_s}$ is an eigenfunction of μ_z, and the corresponding eigenvalue is

$$- \frac{eh}{4\pi m_0 c} (m + 2m_s) . \tag{39.3}$$

As the magnetic quantum number m can take only integer values and the spin quantum number m_s only the values $\pm \frac{1}{2}$, the expression (39.3) is an integer multiple of the Bohr magneton

$$\mu_B = \frac{eh}{4\pi m_0 c} .$$

The eigenfunctions of the total angular momentum, the functions F_{ljm_j} of (38.21), are a combination of the eigenfunctions of two magnetic momenta, namely of $Y_l^m \chi_{+\frac{1}{2}}$ and $Y_l^{m+1} \chi_{-\frac{1}{2}}$. These belong to two different values (to $-\mu_B(m + 1)$ and $-\mu_B m$, respectively) of the z component of the magnetic moment. Consequently, in the F_{ljm_j} state the electron possesses no definite magnetic moment. For electrons moving in atoms this is always the situation, because the angular momentum will be a constant of movement due to the spherical symmetry of the atom, and therefore the state of the electron is described by an angular momentum eigenfunction F_{ljm_j}. However, the quantum-mechanical expectation value of the magnetic moment μ_z can be found. The eigenfunctions $Y_l^m \chi_{+\frac{1}{2}}$ and $Y_l^{m+1} \chi_{-\frac{1}{2}}$ of the magnetic moment figure in the expression for F_{ljm_j} with the coefficients C_+ and C_-, so that the expectation value is

$$\bar{\mu}_z = -\mu_B[| C_+ |^2 (m + 1) + | C_- |^2 m] .$$

Substituting the expressions (38.20) for the coefficients C_+ and C_- into this formula, we find that

$$\bar{\mu}_z = -\mu_B \frac{(l + 1)(2m + 1)}{2l + 1}, \qquad \text{if } j = l + \frac{1}{2},$$

$$\bar{\mu}_z = -\mu_B \frac{l(2m + 1)}{2l + 1}, \qquad \text{if } j = l - \frac{1}{2}. \tag{39.4}$$

It has to be mentioned that this expression for $\bar{\mu}_z$ can be condensed into a single formula if the following abbreviation is introduced:

$$g_L = 1 + \frac{j(j + 1) - l(l + 1) + \frac{3}{4}}{2j(j + 1)} . \tag{39.5}$$

This is called the *Landé g* factor of the electron. Its value is

$$g_L = \frac{2(l+1)}{2l+1}, \qquad \text{if } j = l + \frac{1}{2},$$

$$g_L = \frac{2l}{2l+1}, \qquad \text{if } j = l - \frac{1}{2}.$$

(39.6)

The expectation value of the magnetic moment can now be written in the following simple form

$$\bar{\mu}_z = -\mu_B g_L m_j,$$

(39.7)

where again $m_j = m + \frac{1}{2}$.

The magnetic moment of the protons and neutrons moving in the nuclei can be determined by a similar reasoning. The only important difference is in the expression of the operator of the magnetic moment. The magnetic moment produced by the orbital movement in the case of protons is by analogy to (39.1),

$$\mu_N = + \frac{e}{2M_0 c} \mathbf{N},$$

where M_0 is the mass of the proton. In the case of neutrons no magnetic field is produced by an orbital movement because the neutrons are electrically neutral. Thus for neutrons $\mu_N = 0$.

The two cases can be summed up in a single expression by introducing the charge quantum number τ

$$\mu_N = \frac{\tau e}{2M_0 c} \mathbf{N},$$

(39.8)

where $\tau = 1$ for protons and $\tau = 0$ for neutrons.

The magnetic moment originating from the spin is also more complicated than the expression

$$-\frac{e}{m_0 c} \mathbf{S}$$

valid for electrons. According to measurements the magnetic moment belonging to a proton is $2.793 \, (e/M_0 c) \, \mathbf{S}$, whereas that of a neutron is $-1.913 \, (e/M_0 c) \, \mathbf{S}$. (It has to be emphasized that even the electrically neutral neutron possesses a magnetic field of its own.) That means that the relation between the magnetic spin moment and the spin vector $\mathbf{S}$ is

$$\mu_s = \frac{ge}{2M_0 c}\mathbf{S},$$ (39.9)

where g is the *gyromagnetic ratio*. Its value for protons is $g = 5.585$ and for neutrons $g = -3.827$.

The total magnetic moment is the sum of the orbital and spin angular momenta:

$$\mu = \mu_N + \mu_s = \frac{e}{2M_0 c}(\tau\mathbf{N} + g\mathbf{S}).$$ (39.10)

The eigenvalues of the z component can be determined, similarly to the case of electrons, by the help of the eigenvalue equation:

$$\mu_z(Y_l^m\chi_{m_s}) = \frac{e}{2M_0 c}(\tau N_z + gS_z)(Y_l^m\chi_{m_s}) = \frac{eh}{4\pi M_0 c}(\tau m + gm_s)(Y_l^m\chi_{m_s}).$$

According to this, the expression of the eigenvalues is

$$\mu_z = \mu_M(\tau m + gm_s),$$ (39.11)

where

$$\mu_M = \frac{eh}{4\pi M_0 c} = 5.05 \times 10^{-24} \text{ gauss cm}^3,$$

is the so-called *nuclear magneton*, the nuclear counterpart of the Bohr magneton of the electrons.

Let us determine the expectation value of μ_z in the F_{ljm_j} state. The procedure is quite similar to that used formerly in the case of the electrons. First the eigenvalues (39.11) are written for the states $Y_l^m\chi_{+\frac{1}{2}}$ and $Y_l^{m+1}\chi_{-\frac{1}{2}}$:

$$\mu_M\left(\tau m + \frac{1}{2}g\right), \qquad \mu_M\left(\tau m + \tau - \frac{1}{2}g\right).$$

As the probability of the occurrence of $Y_l^m\chi_{+\frac{1}{2}}$ in the state F_{ljm_j} is $|C_+|^2$, and that of $Y_l^{m+1}\chi_{-\frac{1}{2}}$ is $|C_-|^2$, the expectation value is

$$\bar{\mu}_z = \mu_M\left[|C_+|^2\left(\tau m + \frac{1}{2}g\right) + |C_-|^2\left(\tau m + \tau - \frac{1}{2}g\right)\right].$$

The final expression is reached by inserting the formulae of C_+ and C_-. Introducing again the quantum number $m_j = m + \frac{1}{2}$ and the Landé fac-

121

tor (39.6), we obtain:

$$\bar{\mu}_z = \mu_M[(2 - g_L)\,\tau + (g_L - 1)g]\,m_j\,. \tag{39.12}$$

This can be written in a form analogous to (38.7)

$$\bar{\mu}_z = \mu_M g_s m_j \tag{39.13}$$

where g_s, the Schmidt factor is given by:

$$g_s = (2 - g_L)\,\tau + (g_L - 1)g\,. \tag{39.14}$$

The detailed formulae of g_s are summarized in Table 3.

TABLE 3. *The Schmidt g_s Factor*

Particle	Total angular momentum quantum number	g_s
proton	$j = l + \dfrac{1}{2}$	$g_s = \dfrac{j + 2.29}{j}$
proton	$j = l - \dfrac{1}{2}$	$g_s = \dfrac{j - 1.29}{j + 1}$
neutron	$j = l + \dfrac{1}{2}$	$g_s = -\dfrac{1.91}{j}$
neutron	$j = l - \dfrac{1}{2}$	$g_s = \dfrac{1.91}{j + 1}$

40. The Three-dimensional Potential Well

In the next sections we are concerned with the solution of the radial Schrödinger equation. In case of a three-dimensional potential well with spherical symmetry the potential energy is

$$V(r) = -V_0, \quad \text{if } r < a\,,$$

$$V(r) = 0, \quad \text{if } r > a\,. \tag{40.1}$$

Here V_0 is the depth of the potential well, and a is its radius. For bound states $-V_0 < E < 0$. Then the radial Schrödinger equation (37.17) takes the form :

$$\frac{d^2R}{dr^2} + \frac{2}{r}\frac{dR}{dr} + \left(\alpha^2 - \frac{l(l+1)}{r^2}\right)R = 0, \qquad \text{if } r < a,$$

$$\frac{d^2R}{dr^2} + \frac{2}{r}\frac{dR}{dr} - \left(\kappa^2 + \frac{l(l+1)}{r^2}\right)R = 0, \qquad \text{if } r > a.$$

(40.2)

The notations α and κ given in (33.2) have been introduced here again. Let us first investigate the case of $l = 0$, that is the s-state. Then the differential equations (40.2) are very simple, the general solution follows immediately:

$$R(r) = \frac{1}{r}[b_1 \sin(\alpha r) + c_1 \cos(\alpha r)], \qquad \text{if } r < a,$$

$$R(r) = \frac{1}{r}[b_2 e^{\kappa r} + c_2 e^{-\kappa r}], \qquad \text{if } r > a.$$

As $R(r)$ has to vanish as $r \to \infty$, b_2 has to be equal to zero. Furthermore, $R(r)$ cannot become infinite at $r = 0$, thus also c_1 is zero. The constants b_1 and c_2 have to be chosen in such a way that both R and dR/dr should change continuously at $r = a$. The conditions for continuity are

$$b_1 \sin(\alpha a) = c_2 e^{-\kappa a},$$

$$b_1 \alpha \cos(\alpha a) = -c_2 \kappa e^{-\kappa a}.$$

Dividing the two equations by each other the following eigenvalue condition is reached

$$\kappa = -\alpha \cot(\alpha a). \tag{40.3}$$

This corresponds to the second equation of (33.6), therefore the energy eigenvalues are the same as the energy values corresponding to the odd eigenfunctions in the one-dimensional potential well. Their determination can be performed by the graphical method illustrated in Fig. 13.

The eigenvalue conditions reached when $l \neq 0$ are somewhat more complicated. In this case the solutions of the differential equations (40.2) can be expressed in terms of Bessel functions. The solution for the interior of the potential well for $r < a$ is the following:

$$R(r) = br^{-\frac{1}{2}}J_{l+\frac{1}{2}}(\alpha r), \tag{40.4}$$

when we take into consideration that the solution has to remain finite at $r = 0$.

Taking into consideration that $R(r) \to 0$ as $r \to \infty$ the solution for $r > a$ is

$$R(r) = cr^{-\frac{1}{2}}[J_{l+\frac{1}{2}}(i\kappa r) + iN_{l+\frac{1}{2}}(i\kappa r)], \qquad (40.5)$$

where i is the imaginary unit, and $N_{l+\frac{1}{2}}$ is the Neumann function. The functions (40.4) and (40.5) and their derivatives have to join continuously at $r = a$.

From the continuity conditions a relation between α and κ is obtained, and combining this with (33.13) a set of simultaneous transcendental equations is reached from which α, κ and thus the energy eigenvalues E can be determined.

41. A Particle Locked up in a Sphere with Rigid Walls

A mass point m_0 should be locked up into a sphere with a radius a. Inside the sphere no force acts on the mass point. Thus the potential is

$$V(r) = 0, \qquad \text{if } r < a,$$
$$V(r) = +\infty, \qquad \text{if } r > a. \qquad (41.1)$$

The radial Schrödinger equation inside the sphere is

$$\frac{d^2R}{dr^2} + \frac{2}{r}\frac{dR}{dr} + \left(\frac{8\pi^2 m_0}{h^2}E - \frac{l(l+1)}{r^2}\right)R = 0. \qquad (41.2)$$

The form of this equation is the same as that of the first equation in (40.2), only instead of α the following expression has to be written:

$$\alpha = \sqrt{\frac{8\pi^2 m_0}{h^2}E}. \qquad (41.3)$$

According to (40.4) the solution of $R(r)$ is

$$R(r) = br^{-\frac{1}{2}}J_{l+\frac{1}{2}}(\alpha r). \qquad (41.4)$$

As the mass point cannot escape from the sphere, the wave function has to be equal to zero, both on the surface of the sphere and everywhere outside of it. From the $R(a) = 0$ boundary condition the following equation is

124

reached:

$$J_{l+\frac{1}{2}}(\alpha a) = 0.\tag{41.5}$$

From this transcendental equation α can be determined as well as the energy values E when (41.3) is taken into consideration. For a given angular quantum number l, (41.5) has an infinite number of solutions. For numbering them a radial quantum number n_r can be introduced. The solutions of eq. (41.5) for some l and n_r values are displayed in Table 4. The energy eigenvalues are shown in Fig. 20.

TABLE 4. *Values of αa for the Lowest l and n_r Quantum Numbers*

n_r	$l = 0$	$l = 1$	$l = 2$
1	3.142	4.495	5.764
2	6.284	7.72	9.10
3	9.425	10.90	12.32

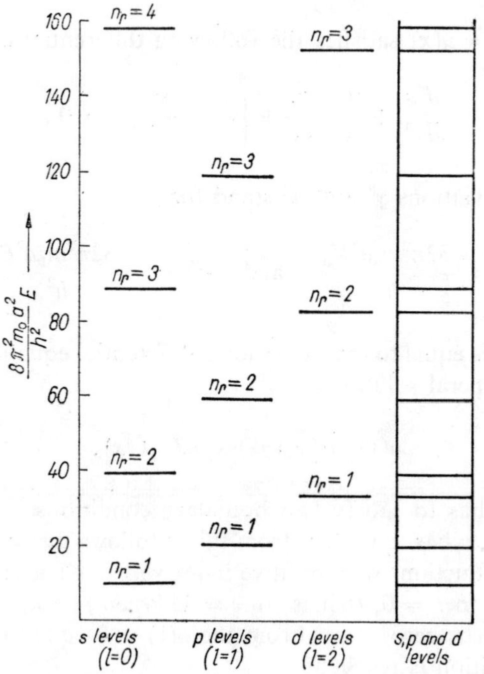

FIG. 20. The energy eigenvalues of a particle in a force-free sphere with infinitely high walls

42. The Exponential Potential

Let us deal with the movement of a mass point m_0 moving in a central field, the potential of which is described by an exponential function:

$$V(r) = -V_0 e^{-r/a}. \tag{42.1}$$

Only the energy of the ground state will be determined, therefore the investigation of the s-state corresponding to $l = 0$ is sufficient. The radial Schrödinger equation is

$$\frac{d^2R}{dr^2} + \frac{2}{r}\frac{dR}{dr} + \frac{8\pi^2 m_0}{h^2}(E + V_0 e^{-r/a})R = 0. \tag{42.2}$$

The form of this equation becomes much simpler if instead of R and r the following new variables are introduced:

$$u = rR(r) \quad \text{and} \quad x = e^{-r/2a}. \tag{42.3}$$

The function $u = u(x)$ satisfies the following differential equation:

$$\frac{d^2u}{dx^2} + \frac{1}{x}\frac{du}{dx} + \left(\gamma^2 - \frac{\kappa^2}{x^2}\right)u = 0, \tag{42.4}$$

where the abbreviations γ^2 and κ^2 stand for

$$\gamma^2 = \frac{32\pi^2 m_0 a^2 V_0}{h^2} \quad \text{and} \quad \kappa^2 = -\frac{32\pi^2 m_0 a^2 E}{h^2}. \tag{42.5}$$

Equation (42.4) is equal to the well-known differential equation of the Bessel functions. Its general solution is

$$u(x) = bJ_\kappa(\gamma x) + cJ_{-\kappa}(\gamma x).$$

The solution has to satisfy two boundary conditions. First, as $r \to \infty$, that is, as $x \to 0$, u has to vanish; from this if follows that $c = 0$. (At $x \to 0$ only the Bessel functions with positive index vanish.) The second boundary condition refers to $r = 0$, that is to $x = 1$: when $r = 0$, $R(r)$ stays finite, so that u has to be zero there. From the $u(1) = 0$ equation the following eigenvalue condition is reached:

$$J_\kappa(\gamma) = 0. \tag{42.6}$$

126

The γ quantity is determined by the potential V_0 and the parameters a, the κ belonging to value γ can be calculated from (42.6). Then the eigenvalue E can be easily determined from the defining expression of κ.

Let us apply our results to the investigation of the deuteron, the bound system consisting of a proton and a neutron. In this case the reduced mass of the proton–neutron system has to be substituted in place of the mass m_0:*

$$m_0 = \frac{M_n M_p}{M_n + M_p} = 0.8368 \times 10^{-24} \text{ g},\tag{42.7}$$

and a is the range of the nuclear forces, for which the following value is taken:

$$a = 2.18 \times 10^{-13} \text{ cm}.\tag{42.8}$$

The experimental value of the binding energy E of the deuteron is

$$E = -2.23 \text{ MeV}.\tag{42.9}$$

Substituting these data into (42.5), the following numerical values are obtained:

$$\kappa = 0.998 \quad \text{and} \quad V_0 = 2.24 \, \gamma^2 \text{ MeV}.\tag{42.10}$$

The solution of the transcendental equation (42.6) for this value of κ is $\gamma = 3.82$, this yields a value for V_0

$$V_0 = 32.7 \text{ MeV}.\tag{42.11}$$

This means that the depth V_0 of the attractive potential well is an order of magnitude higher than the binding energy E. The kinetic energy counterbalances almost completely the potential energy. This relatively high kinetic energy is due to the Heisenberg relation.

43. The Three-dimensional Oscillator

The potential of the three-dimensional oscillator can be put on analogy with (32.3), but instead of the distance x the distance r measured from the centre of force is used. Thus the potential is

$$V(r) = 2\pi^2 m_0 v^2 r^2.\tag{43.1}$$

* This will be dealt with in Section 47.

Substituting this into the Schrödinger equation, we obtain:

$$\frac{d^2R}{dr^2} + \frac{2}{r}\frac{dR}{dr} + \left[\frac{8\pi^2 m_0 E}{h^2} - \frac{16\pi^4 m_0^2 v^2}{h^2}r^2 - \frac{l(l+1)}{r^2}\right]R = 0. \quad (43.2)$$

Let us introduce again the abbreviations of (32.5) and express R as

$$R(r) = r^l e^{-\frac{1}{2}\alpha r^2} v(r). \quad (43.3)$$

After these substitutions the differential equation (43.2) will take the following form:

$$\frac{d^2v}{dr^2} + 2\left(\frac{l+1}{r} - \alpha r\right)\frac{dv}{dr} - \left[2\alpha\left(l + \frac{3}{2}\right) - \lambda\right]v = 0. \quad (43.4)$$

Changing from the variable r to the new variable $t = \alpha r^2$ and denoting for the sake of simplicity differentiation with respect to t by a prime, we obtain

$$tv''(t) + \left(l + \frac{3}{2} - t\right)v'(t) - \frac{1}{2}\left(l + \frac{3}{2} - \frac{\lambda}{2\alpha}\right)v(t) = 0. \quad (43.5)$$

This equation can be solved by the polynomial method described in Section 32. v is put in the form of a polynomial of degree n_r:

$$v(t) = \sum_{p=0}^{n_r} a_p t^p. \quad (43.6)$$

Substituting this in (43.5) and arranging this new equation according to increasing powers of t, we get

$$\sum_{p=0}^{n_r-1}\left[(p+1)\left(p+l+\frac{3}{2}\right)a_{p+1} - \frac{1}{2}\left(2p+l+\frac{3}{2} - \frac{\lambda}{2\alpha}\right)a_p\right]t^p$$

$$- \frac{1}{2}\left[2n_r + l + \frac{3}{2} - \frac{\lambda}{2\alpha}\right]a_{n_r}t^{n_r} = 0. \quad (43.7)$$

This equation holds for every value of t if the coefficients of the different powers of t are independently equal to zero. For $0 \leq p \leq n_r - 1$ the coefficient of t^p is equal to zero, if

$$a_{p+1} = \frac{2p + l + 3/2 - \lambda/2\alpha}{2(p+1)(p+l+3/2)}a_p. \quad (43.8)$$

128

This formula makes the successive determination of the coefficients of the polynomials possible. (a_0 can be chosen *ad libitum*, or it can be determined from the normalization condition.) As in (43.7) also the coefficient of t^{n_r} has to be equal to zero,

$$\frac{\lambda}{2\alpha} = 2n_r + l + \frac{3}{2}. \tag{43.9}$$

That means that eq. (43.5) can be solved by a finite polynomial only if the quantity $\lambda/2\alpha$ figuring in the equation is a number having the form shown in (43.9), where l is the orbital quantum number and n_r is also a nonnegative integer, called the *radial quantum number*. Substituting the expressions for λ and α in expression (43.9), we obtain:

$$E = \left(2n_r + l + \frac{3}{2}\right) hv. \tag{43.10}$$

By introducing the principal quantum number n, through the definition

$$n = 2n_r + l \tag{43.11}$$

the expression for E gets the following form:

$$E = \left(n + \frac{3}{2}\right) hv. \tag{43.12}$$

The energy of the ground state is obtained for $n = 0$: $E = \frac{3}{2} hv$. This is three times higher than in the case of the linear oscillator. The excitation energy of the nth level is

$$E_n - E_0 = nhv,$$

an integer multiple of hv. The principal quantum number n introduced in (43.11) can take on every even or odd value, depending on whether l is even or odd. The quantum numbers n_r and l can combine in various ways for a given n value, thus several eigenfunctions belong to an energy. (The energy levels are degenerate.) If n is an even number then l can take any even value between 0 and n, their number is $n/2 + 1$. To every possible l value still $2l + 1$ different eigenfunctions belong in accordance with the $0, \pm 1, \ldots \pm l$ values of the magnetic quantum number. Therefore the number of eigenfunctions belonging to an even n is

$$\sum_0^n (2l + 1) = (n + 1)(n + 2)/2.$$

(The summation is to be performed only over the even values of l!) A similar result is reached if n is an odd number. In this case l can take odd numbers between 1 and n, and the total number of the eigenfunctions is

$$\sum_{1}^{n} (2l + 1) = (n + 1) (n + 2)/2 \,.$$

As an example let us investigate in detail the $n = 3$ case. According to rule (43.11) we have two possibilities: either $n_r = 0$ and $l = 3$, or $n_r = 1$ and $l = 1$. In the first case the degree of the v polynomial is zero, thus $v = 1$. According to (43.3) the radial wave function is: $R = r^3 e^{-\alpha(r^2/2)}$. The total ψ wave function is obtained if R is multiplied by the spherical functions Y_l^m. For $l = 3$ seven such spherical functions exist, their form has already been presented in Section 37. In the second case, when $n_r = l = 1$, the coefficients of the v polynomial can be determined from recursion formula (43.8). We obtain

$$v = 1 - \frac{2}{5} t = 1 - \frac{2}{5} \alpha r^2 \,.$$

And the radial wave function is:

$$R(r) = r \left(1 - \frac{2}{5} \alpha r^2 \right) e^{-\alpha(r^2/2)} \,.$$

To obtain the total wave function ψ this has still to be multiplied by one of the spherical functions Y_l^m. The number of the spherical functions is now three. The ten ψ functions obtained in this way and expressed in terms of Cartesian coordinates x, y and z are summarized in Table 5. For the sake of simplicity the normalizing factors have been omitted.

TABLE 5. *The Eigenfunctions of the Spherical Oscillator for* $n = 3$

n_r	l	m	ψ
0	3	0	$e^{-\alpha(r^2/2)} z(2z^2 - 3x^2 - 3y^2)$
0	3	± 1	$e^{-\alpha(r^2/2)} (x \pm iy) (4z^2 - x^2 - y^2)$
0	3	± 2	$e^{-\alpha(r^2/2)} (x \pm iy)^2 z$
0	3	± 3	$e^{-\alpha(r^2/2)} (x \pm iy)^3$
1	1	0	$e^{-\alpha(r^2/2)} \left(1 - \frac{2}{5} \alpha r^2 \right) z$
1	1	± 1	$e^{-\alpha(r^2/2)} \left(1 - \frac{2}{5} \alpha r^2 \right) (x \pm iy)$

Each of these ten eigenfunctions belongs to the eigenvalue $E_3 = hv(3 + 3/2)$. Naturally also all arbitrary linear combinations of these functions belong to the same energy value.

In the above-mentioned theory of the spherical oscillator we made use of the spherical symmetry of the potential and the problem was solved in the polar space-coordinate system. There is also another possibility for the solution, where the Schrödinger equation is written in the rectangular Cartesian coordinate system. After introducing the quantities α and λ in the same way as before we find the Schrödinger equation to be:

$$\frac{\partial^2 \psi}{\partial x^2} + \frac{\partial^2 \psi}{\partial y^2} + \frac{\partial^2 \psi}{\partial z^2} + [\lambda - \alpha(x^2 + y^2 + z^2)] \psi = 0 . \tag{43.13}$$

Let ψ be

$$\psi(x, y, z) = f_1(x) f_2(y) f_3(z) \tag{43.14}$$

and substitute this into (43.13). After dividing by ψ we find that

$$\frac{1}{f_1(x)} \left(\frac{d^2 f_1}{dx^2} - \alpha x^2 f_1(x) \right) + \frac{1}{f_2(y)} \left(\frac{d^2 f_2}{dy^2} - \alpha y^2 f_2(y) \right)$$

$$+ \frac{1}{f_3(z)} \left(\frac{d^2 f_3}{dz^2} - \alpha z^2 f_3(z) \right) + \lambda = 0 .$$

This equation can hold for every value of the variables x, y and z only if each single terms is constant. Thus,

$$\frac{d^2 f_1}{dx^2} + (\lambda_1 - \alpha x^2) f_1(x) = 0 ,$$

$$\frac{d^2 f_2}{dy^2} + (\lambda_2 - \alpha y^2) f_2(y) = 0 , \tag{43.15}$$

$$\frac{d^2 f_3}{dz^2} + (\lambda_3 - \alpha z^2) f_3(z) = 0 ,$$

where

$$\lambda_1 + \lambda_2 + \lambda_3 = \lambda . \tag{43.16}$$

Each of the eigenvalue equations (43.15) shows the same structure as eq. (32.6), the equation of the linear oscillator. Thus their solutions follow immediately from (32.21) and (32.25):

$$\lambda_1 = (2n_1 + 1) \alpha , \quad f_1(x) = A_{n1} e^{-\alpha(x^2/2)} H_{n1}(\sqrt{\alpha}\, x) ,$$

$$\lambda_2 = (2n_2 + 1) \alpha , \quad f_2(y) = A_{n2} e^{-\alpha(y^2/2)} H_{n2}(\sqrt{\alpha}\, y) ,$$

$$\lambda_3 = (2n_3 + 1) \alpha , \quad f_3(z) = A_{n3} e^{-\alpha(z^2/2)} H_{n3}(\sqrt{\alpha}\, z) .$$

The energy E and wave function ψ of the oscillator can be determined according to (43.16) and (43.14):

$$E = hv \left(n_1 + n_2 + n_3 + \frac{3}{2}\right) = hv \left(n + \frac{3}{2}\right), \qquad (43.17)$$

$$\psi = A e^{-\alpha(r^2/2)} H_{n1}(\sqrt{\alpha}\, x)\, H_{n2}(\sqrt{\alpha}\, y)\, H_{n3}(\sqrt{\alpha}\, z). \qquad (43.18)$$

The expression obtained for the energy equals (43.12), only tne principal quantum number has now the form $n = n_1 + n_2 + n_3$. The number of the wave functions belonging to a given principal quantum number n is equal to the number that shows in how many ways the number n can be decomposed into the non-negative integers n_1, n_2, and n_3, so that it is

$$\frac{(n + 1)(n + 2)}{2}.$$

(The same result was reached from the polar coordinate treatment of the problem.) The eigenfunctions belonging to the $n = 3$ principle quantum number are summed up in Table 6. (The normalizing coefficients are also here omitted.)

The eigenfunctions ψ of Tables 5 and 6 differ from each other. This means, for instance, that the functions of Table 6 are eigenfunctions of the energy operator of the oscillator, but they are not eigenfunctions of the orbital angular momentum. The eigenfunctions of both the energy and the angular

TABLE 6. *The Eigenfunctions of the Spherical Oscillator for $n = 3$*

n_1	n_2	n_3	ψ
0	0	3	$e^{-\alpha(r^2/2)}\, z(2\alpha z^2 - 3)$
0	1	2	$e^{-\alpha(r^2/2)}\, y(2\alpha z^2 - 1)$
0	2	1	$e^{-\alpha(r^2/2)}\, z(2\alpha y^2 - 1)$
0	3	0	$e^{-\alpha(r^2/2)}\, y(2\alpha y^2 - 3)$
1	0	2	$e^{-\alpha(r^2/2)}\, x(2\alpha z^2 - 1)$
1	1	1	$e^{-\alpha(r^2/2)}\, xyz$
1	2	0	$e^{-\alpha(r^2/2)}\, x(2\alpha y^2 - 1)$
2	0	1	$e^{-\alpha(r^2/2)}\, z(2\alpha x^2 - 1)$
2	1	0	$e^{-\alpha(r^2/2)}\, y(2\alpha x^2 - 1)$
3	0	0	$e^{-\alpha(r^2/2)}\, x(2\alpha x^2 - 3)$

momentum are presented in Table 5. By linear combinations the functions of the two tables can, of course, be produced from each other. For example, the solution corresponding to the quantum numbers $n_r = 1$, $l = 1$ and $m = = 0$ can be expressed by the functions $\psi_{n_1 n_2 n_3}$ of Table 6 in the following way:

$$\psi_{(n_r=1,\ l=1,\ m=0)} = -\frac{1}{5}(\psi_{003} + \psi_{021} + \psi_{201}).$$

44. The Hydrogen Atom

It is well known that the H-atom consists of a nucleus (proton) and an electron. In the following the nucleus will be regarded as stationary. This is equivalent to taking the mass of the nucleus infinitely large compared to that of the electron. Let us choose the charge of the nucleus to be Ze, then we can deal with the H-atom and the He^+, Li^{++}, Be^{+++}, ... ions in one step by simply choosing Z equal to 1, 2, 3, ...

For the potential energy we have:

$$V(r) = +\frac{Ze}{r}(-e) = -\frac{Ze^2}{r}, \tag{44.1}$$

where $-e$ is the charge of the electron, r its distance from the nucleus. The radial Schrödinger equation has the form

$$\frac{d^2R}{dr^2} + \frac{2}{r}\frac{dR}{dr} + \left(\frac{8\pi^2 m_0}{h^2}E + \frac{8\pi^2 m_0}{h^2}\frac{Ze^2}{r} - \frac{l(l+1)}{r^2}\right)R = 0. \tag{44.2}$$

Let us introduce the following simplifying notation:

$$A = \frac{8\pi^2 m_0}{h^2}E, \quad B = \frac{4\pi^2 m_0 Ze^2}{h^2}, \quad C = -l(l+1). \tag{44.3}$$

Then

$$\frac{d^2R}{dr^2} + \frac{2}{r}\frac{dR}{dr} + \left(A + \frac{2B}{r} + \frac{C}{r^2}\right)R = 0. \tag{44.4}$$

The solutions differ considerably from each other for positive and negative E, i.e. A, values. For large r values this is obvious at first sight from the form of the equation. For large r values the form of the equation is namely the following:

$$\frac{d^2R}{dr^2} + AR = 0. \tag{44.5}$$

The function R is an oscillating or exponentially decaying function of r depending on whether A is positive or negative. This means an important difference. In the following we are going to deal in detail with the bound states, thus with the case of negative E, i.e. A, values.

Let A be written as

$$A = - \frac{1}{r_0^2},\qquad (44.6)$$

where r_0 is some length. This corresponds to the fact that the dimension of A is [length]$^{-2}$, as can be seen directly from the equation, too. It follows from the equation that the dimension of A is the same as that of C/r^2, and as C is a dimensionless number, it follows that the dimension of C/r^2 and thus that of A is [length]$^{-2}$. Let us introduce the dimensionless quantity ρ as

$$\rho = \frac{2r}{r_0}.\qquad (44.7)$$

Using these formulae one gets from the differential equation of R

$$\left(\frac{2}{r_0}\right)^2 \frac{d^2R}{d\rho^2} + \left(\frac{2}{r_0}\right)^2 \frac{2}{\rho}\frac{dR}{d\rho} + \left[-\frac{1}{r_0^2} + \frac{2}{r_0}\frac{2B}{\rho} + \left(\frac{2}{r_0}\right)^2 \frac{C}{\rho^2}\right]R = 0,$$

$$\frac{d^2R}{d\rho^2} + \frac{2}{\rho}\frac{dR}{d\rho} + \left[-\frac{1}{4} + \frac{r_0 B}{\rho} - \frac{l(l+1)}{\rho^2}\right]R = 0.\qquad (44.8)$$

From this it follows that the form of the equation valid for $r = \infty$, i.e. $\rho = \infty$, is

$$\frac{d^2R}{d\rho^2} = \frac{1}{4}R.\qquad (44.9)$$

Apart from a constant the solution of this is $R = e^{\pm\rho/2}$.

As the eigenfunction has to vanish at infinity, only the negative exponent comes into question, thus $R = e^{-\rho/2}$.

The solution of the equation valid for finite r, i.e. ρ values, is put as

$$R = e^{-\rho/2} u(\rho),\qquad (44.10)$$

and from this we get

$$\frac{dR}{d\rho} = \left(-\frac{1}{2}u + \frac{du}{d\rho}\right)e^{-\rho/2}, \qquad \frac{d^2R}{d\rho^2} = \left(\frac{1}{4}u - \frac{du}{d\rho} + \frac{d^2u}{d\rho^2}\right)e^{-\rho/2}.$$

134

Substituting these expressions into the former equation and dividing it by $e^{-\rho/2}$, we find that

$$\frac{d^2u}{d\rho^2} + \left(\frac{2}{\rho} - 1\right)\frac{du}{d\rho} + \left[(r_0B - 1)\frac{1}{\rho} - \frac{l(l+1)}{\rho^2}\right]u = 0. \quad (44.11)$$

This is a differential equation for u, with a singularity at $\rho = 0$. A series solution of this is assumed. From the theory of differential equations it follows that in this case, as $\rho = 0$ is a pole, u can be represented by such a series where the term of lowest degree is, apart from a constant factor, equal to ρ^l. Thus u is chosen to be

$$u = \rho^l w(\rho) = \rho^l(a_0 + a_1\rho + a_2\rho^2 + \ldots),$$

that is

$$u = \sum_{p=1}^{\infty} a_p\rho^{p+l}. \quad (44.12)$$

Let us substitute u into the differential equation for u. For this first the following quantities should be calculated:

$$\frac{u}{\rho}, \quad \frac{u}{\rho^2}, \quad \frac{du}{d\rho}, \quad \frac{d^2u}{d\rho^2} \quad \text{and} \quad \frac{1}{\rho}\frac{du}{d\rho}:$$

$$\frac{u}{\rho} = \sum_{p=0}^{\infty} a_p\rho^{p+l-1},$$

$$\frac{u}{\rho^2} = \sum_{p=0}^{\infty} a_p\rho^{p+l-2},$$

$$\frac{du}{d\rho} = \sum_{p=0}^{\infty} (p+l) a_p\rho^{p+l-1},$$

$$\frac{d^2u}{d\rho^2} = \sum_{p=0}^{\infty} (p+l)(p+l-1) a_p\rho^{p+l-2},$$

$$\frac{1}{\rho}\frac{du}{d\rho} = \sum_{p=0}^{\infty} (p+l) a_p\rho^{p+l-2}.$$

The result of the substitution after rearranging is

$$\sum_{p=0}^{\infty} \left[(p+l+1)(p+l) a_{p+1} + 2(p+l+1) a_{p+1} - (p+l) a_p + (r_0B - 1) a_p\right.$$
$$\left. - l(l+1) a_{p+1}\right] \rho^{p+l-1} = 0. \quad (44.13)$$

This can hold for every ρ only if the coefficients of each of the powers of ρ vanish so that

$$(p + l + 1)(p + l)\, a_{p+1} + 2(p + l + 1)\, a_{p+1} - (p + l)\, a_p +$$
$$+ (r_0 B - 1)\, a_p - l(l + 1)\, a_{p+1} = 0 . \qquad (44.14)$$

Rearranging this formula the following recurrence relation is obtained to determine the coefficients a_p:

$$[(p + l + 1)(p + l + 2) - l(l + 1)]\, a_{p+1} = (p + l + 1 - r_0 B)\, a_p . \quad (44.15)$$

Apart from a constant coefficient this equation determines fully the coefficients a_p, that is u. We can make sure that R vanishes at infinity by preventing p becoming infinite, by requiring that u should be a polynomial. (That this condition is not only sufficient but also necessary is not going to be proved here.) As seen from the above recurrence relation, $a_{p+1}, a_{p+2}, a_{p+3}, \ldots$ all vanish if $a_p = 0$. Thus if we want $a_{n_r+1}, a_{n_r+2}, \ldots$ to vanish, then it is required that

$$n_r + l + 1 - r_0 B = 0 , \qquad (44.16)$$

that is

$$r_0 B = n_r + l + 1 = n . \qquad (44.17)$$

And thus

$$r_0^2 B^2 = -\frac{B^2}{A} = (n_r + l + 1)^2 = n^2 ,$$

$$-\frac{\left(\dfrac{4\pi^2 m_0 Z e^2}{h^2}\right)^2}{\dfrac{8\pi^2 m_0 E}{h^2}} = -\frac{2\pi^2 m_0 Z^2 e^4}{h^2 E} = (n_r + l + 1)^2 = n^2 .$$

By writing E_n instead of E we find

$$E_n = -\frac{2\pi^2 m_0 Z^2 e^4}{h^2} \cdot \frac{1}{(n_r + l + 1)^2} = -\frac{2\pi^2 m_0 Z^2 e^4}{h^2} \cdot \frac{1}{n^2} . \quad (44.18)$$

These E_n values are for $Z = 1$ just the energy levels of H.

Before the detailed discussion of the solution first R has to be dealt with. We have already seen that

$$R = e^{-\rho/2} \rho^l w(\rho) = e^{-\rho/2} u(\rho) ,$$

$$u(\rho) = \rho^l w(\rho) .$$

Substituting this form of u in the differential equation of u, and taking into consideration that $r_0B = n$, we find for w a differential equation that can be brought by simple calculations into the following form:

$$\rho \frac{d^2w}{d\rho^2} + [2(l + 1) - \rho] \frac{dw}{d\rho} + (n + l + 1) w = 0 . \qquad (44.19)$$

This is the differential equation of the $(2l + 1)$st derivative of the $(n + l)$th Laguerre polynomial, thus

$$L_k(\rho) = e^\rho \frac{d^k}{d\rho^k} (\rho^k e^{-\rho}) . \qquad (44.20)$$

As mentioned, w is the $(2l + 1)$st derivative of the $(n + l)$th Laguerre polynomial. Thus,

$$w = \frac{d^{2l+1}}{d\rho^{2l+1}} L_{n+l}(\rho) = \frac{d^{2l+1}}{d\rho^{2l+1}} \left[e^\rho \frac{d^{n+l}}{d\rho^{n+l}} (\rho^{n+l} e^{-\rho}) \right] . \qquad (44.21)$$

This is denoted by $L_{n+l}^{(2l+1)}$. This leads to the result

$$R = R_{nl} = e^{-\rho/2} \rho^l L_{n+l}^{(2l+1)} (\rho) . \qquad (44.22)$$

Moreover, the total eigenfunction is

$$\psi_{nlm} = N e^{-\rho/2} \rho^l L_{n+l}^{(2l+1)}(\rho) Y_l^m(\vartheta, \phi) , \qquad (44.23)$$

where N is a constant, the so-called *normalizing constant*, determinable from the normalizing conditions. For N we obtain

$$N = \sqrt{\left(\frac{2Z}{na_H} \right)^3 \frac{(n - l - 1)!}{2n[(n+l)!]^3}} , \qquad (44.24)$$

where

$$a_H = \frac{h^2}{4\pi^2 m_0 e^2} ,$$

that is, a_H is the first Bohr radius of H, used also in quantum mechanics as the atomic unit of length.

Let us now determine the number of different eigenfunctions associated with a given eigenvalue, i.e. with a fixed value of n.

It has already been shown that

$$n = n_r + l + 1 ,$$

so that the maximum value of l for a given n is $n - 1$, its minimum value is zero. Thus in case of a given n, l can take n different values. For every l we have the following m values : $-l, -(l - 1), \ldots -3, -2, -1, 0, +1, +2, +3, \ldots, +(l - 1), +l$: altogether $2l + 1$ values. This obviously implies that the number of different, that is linearly independent, eigenfunctions associated with a given E_n, and n, respectively, is

$$\sum_{l=0}^{n-1} (2l + 1) = [1 + 2(n - 1) + 1]\frac{n}{2} = n^2. \qquad (44.25)$$

Thus every state of energy E_n is $n^2 - 1$ times degenerate.

It has to be mentioned that these calculations do not contain the spin; if this also is considered and the fine structure of the energy levels is disregarded then $2n^2$ different eigenfunctions are associated with each E_n energy level.

Now we turn to the discussion of the solutions corresponding to the negative energy states. First the quantum numbers and their meaning will be dealt with.

The energy E_n depends only on n, thus n is the principal quantum number. Earlier it was shown that

$$n = n_r + l + 1.$$

For this it is obvious that in Bohr's theory n_r corresponds to the radial, l to the orbital quantum number. The smallest value of l is zero, its highest value, for a given n, is $n - 1$. This means that in case of a given n, l can take values $0, 1, 2, \ldots, n - 1$. The orbital quantum number has already been introduced in Bohr's theory but we could not give a theoretical basis for it. l was introduced in the Bohr theory in the following way: the azimuthal quantum number k was obtained from the second quantum condition. Its possible values were $1, 2, 3, \ldots, n$. The value of k was reduced by 1, referring simply to different empirical findings. This resulted in the quantum number $l = k - 1$. Now, however, the values of l being in agreement with experiment are obtained directly from the theory.

Let us deal now with the quantum number n_r. The polynomial

$$L_{n+l}^{(2l+1)}(\rho)$$

is contained in that part of the eigenfunction which depends on r, or ρ. The degree of this polynomial is

$$n + l - (2l + 1) = n - l - 1 = n_r. \qquad (44.26)$$

Such a polynomial has generally n_r zeros. In case of the polynomials $L_{n+l}^{(2l+1)}$ it can be proved that none of their zeros coincides with another one. Thus the number of spherical surfaces where R, or ψ, corresponding to the quantum numbers n, l, are identically equal to zero, is n_r. That means that n_r, i.e. the old radial quantum number, gives the number of those spherical surfaces where ψ is identically equal to zero. Or when passing over to the time-dependent eigenfunctions $\Psi = \psi e^{-(2\pi i/h)\,E_{n^l}}$, we may say that n_r is the number of the nodal points of the Ψ-wave. The other quantum numbers have a similar meaning.

The ϑ dependence of the eigenfunctions is given by P_l^m, which is a polynomial of degree l in $\cos \vartheta$ and $\sin \vartheta$. The nodal surfaces are here cones, and their common axis is the z-axis. The number of cones is l. The ϕ-dependence of the eigenfunctions is given by $\cos m\phi$ and $\sin m\phi$. m is thus the number of nodal planes through the z-axis.

In the following we are briefly concerned with the interpretation of the wave mechanical results obtained for the H-atom. As we saw during the discussion, the model-like concept of electron orbits of the Bohr theory did not appear anywhere. Here only the probability of the presence of the electron in a given volume element can be given. It is known that the probability that the electron is present in a volume element dv is $\psi^*\psi\, dv$. In the case of a stationary state a specially clear picture of the r dependence of this probability is gained, when the $\psi^*\psi\, dv$ expression is integrated over ϑ and ϕ. This yields the expression $R_{nl}^2 r^2 dr$. Figure 21 shows the $R_{nl}^2 r^2$ radial distributions for several states. All these distributions have a primary maximum. It is interesting that the r values corresponding to the places of maximum probability agree more or less with the radii of the corresponding spherical Bohr orbits. $R_{nl}^2 r^2$ gives only the r dependence of the probability distribution. The probability that the electron can be found around the nucleus in a dv volume element is given by the following formula:

$$\psi^*\psi dv = \left(\frac{2Z}{na_H}\right)^3 \frac{(n-l-1)!}{2n[(n+l)!]^3} \frac{2l+1}{4\pi} \frac{(l-m)!}{(l+m)!}$$

$$\times \left[\left(\frac{2Zr}{na_H}\right)^l L_{n+l}^{(2l+1)}\left(\frac{2Zr}{na_H}\right) P_l^m(\cos \vartheta)\right]^2 dv \,.$$

This expression depends also on ϑ, but is independent of ϕ, that is, it yields a distribution symmetrical around the z-axis.

Finally we mentioned that our treatment of the H-atom was somewhat approximate, because it did not account for the fine-structure of the H-spectrum and for the spin. The fine-structure and the spin are explained by the so-called Dirac theory, the relativistic quantum mechanics, which is

139

based on the Dirac equation. This is, in contrast to the Schrödinger equation, relativistically invariant, and can be regarded as the generalization of the Schrödinger equation.

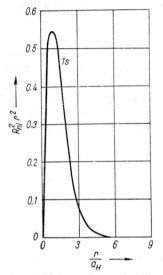

FIG. 21a. The radial probability distribution of the 1s electron of the H-atom

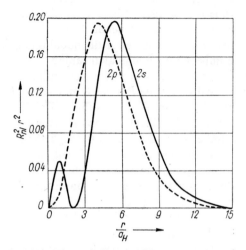

FIG. 21b. The radial probability distribution of the 2s and 2p electrons of the H-atom

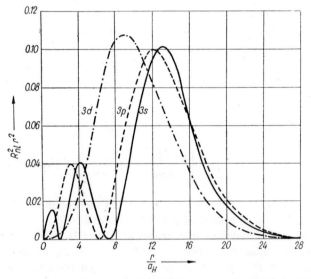

FIG. 21c. The radial probability distribution of the 3s, 3p and 3d electrons of the H-atom

45. The Kratzer Potential

Let us solve the Schrödinger equation for the case of the following, so-called Kratzer potential:

$$V(r) = -\frac{C}{r} + \frac{D}{r^2}, \tag{45.1}$$

where C and D are positive constants. The interaction between the two atoms of a diatomic molecule can be described, with a good approximation, by such a potential. The potential curve is illustrated in Fig. 22. The lowest point of the curve corresponds to the equilibrium position of the molecule. The following simplifying symbols are introduced into the radial Schrödinger equation:

$$u = rR(r), \qquad \rho = \frac{C}{2D} r,$$

$$\gamma^2 = \frac{8\pi^2 m_0 D}{h^2}, \tag{45.2}$$

$$\kappa^2 = -\frac{32\pi^2 m_0 D^2}{h^2 C^2} E.$$

This yields the following equation:

$$\frac{d^2u}{d\rho^2} + \left(-\kappa^2 + \frac{2\gamma^2}{\rho} \right.$$

$$\left. -\frac{\gamma^2 + l(l+1)}{\rho^2} \right) u = 0. \tag{45.3}$$

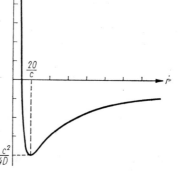

FIG. 22. The Kratzer potential

The solution of this equation is similar to the procedure for the case of the oscillator or the H-atom. As $\rho \to \infty$ the asymptotic form of function u is obviously $u = e^{-\kappa\rho}$. And when $\rho \to 0$, if the solution is required in the form of $u = \rho^\lambda$, from (45.3) the following equation is found for index λ:

$$\lambda(\lambda - 1) = \gamma^2 + l(l+1).$$

This is a quadratic equation for λ, which has, however, only one positive root:

$$\lambda = \frac{1}{2} + \sqrt{\gamma^2 + \left(l + \frac{1}{2} \right)^2}. \tag{45.4}$$

It is reasonable to require the exact solution of (45.3) in the following form:

$$u(\rho) = \rho^\lambda e^{-\kappa\rho} v(\rho),$$ (45.5)

where $v(\rho)$ is a polynomial. The substitution of this expression in (45.3) yields

$$\rho \frac{d^2v}{d\rho^2} + 2(\lambda - \kappa\rho)\frac{dv}{d\rho} + 2(\gamma^2 - \lambda\kappa) v = 0.$$ (45.6)

The structure of this equation is very similar to that of eq. (43.5) discussed in connection with the problem of the rigid oscillator. In fact, by introducing the variable $t = 2\kappa\rho$, (45.6) gets the form

$$t \frac{d^2v}{dt^2} + (2\lambda - t)\frac{dv}{dt} + \left(\frac{\gamma^2}{\kappa} - \lambda\right) v(t) = 0.$$ (45.7)

The structure of this equation is the same as that of (43.5). As discussed in Section 43, this equation has a polynomial solution if in the last term the coefficient of v is a non-negative integer, n_r (n_r will be the degree of the polynomial). In our present eigenvalue condition case

$$\frac{\gamma^2}{\kappa} - \lambda = n_r,$$

or

$$\kappa = \frac{\gamma^2}{n_r + \lambda}.$$ (45.8)

From this the energy eigenvalue E is

$$E = -\frac{2\pi^2 m_0 C^2}{h^2} \cdot \frac{\kappa^2}{\gamma^4} = -\frac{2\pi^2 m_0 C^2}{h^2} \cdot \frac{1}{\left[n_r + \frac{1}{2} + \sqrt{\gamma^2 + \left(l + \frac{1}{2}\right)^2}\right]^2}.$$ (45.9)

It is worth investigating the energy levels for $\gamma \ll 1$ and $\gamma \gg 1$. In the first case let us expand E into a series in powers of γ^2. Writing in full the first two terms of the power series we find that

$$E = -\frac{2\pi^2 m_0 C^2}{h^2} \frac{1}{n^2}\left\{1 - \frac{\gamma^2}{n(2l + 1)} + \dots\right\},$$ (45.10)

where $n = n_r + l + 1$. For the limiting case of $\gamma = 0$ the energy levels correspond to those of the H-atom (see eq. (44.18)). This is obvious as the

142

Kratzer potential (45.1) changes into the Coulomb potential (see eq. (44.1)) by the substitutions $D = 0$ and $C = Ze^2$.

Let us now investigate the $\gamma \gg 1$ case. Now E can be expanded into a series in powers of $1/\gamma$.

$$E = \frac{C^2}{4D}\left\{-1 + \frac{2\left(n_r + \frac{1}{2}\right)}{\gamma} + \frac{\left(l + \frac{1}{2}\right)^2}{\gamma^2} - \frac{3\left(n_r + \frac{1}{2}\right)^2}{\gamma^2}\right.$$
$$\left. - \frac{3\left(n_r + \frac{1}{2}\right)\left(l + \frac{1}{2}\right)^2}{\gamma^3} + \ldots\right\}. \tag{45.11}$$

For $\gamma \to \infty$, $E \to - C^2/4D$. As seen from Fig. 22 this value just corresponds to the lowest point of the potential curve. At a fixed value of the constant D of the potential the $\gamma \to \infty$ limiting case can be reached if in (45.2) we let the Planck constant (h) tend towards zero. That means that in the classical limiting case of $h \to 0$ the binding energy of the molecule is equal to the value of the potential corresponding to the equilibrium position. The quantum-mechanical value corresponding to $h \neq 0$ differs from this due to the fact that according to Heisenberg's relation the molecule cannot be in the equilibrium state but oscillates in the environment of this state. For investigating the oscillations let us expand the potential (45.1) into a series around the equilibrium positions ($r_0 = 2D/C$):

$$V(r) = - \frac{C^2}{4D} + \frac{C^2}{4Dr_0^2}(r - r_0)^2 .$$

We see that an oscillator potential is added to the value $-C^2/4D$ corresponding to the lowest point. The classical frequency of this potential according to (32.1) is

$$\nu = \frac{1}{2\pi}\sqrt{\frac{C^2}{2Dr_0^2 m_0}} . \tag{45.12}$$

Instead of the constants C and D of the Kratzer potential the above determined frequency ν as well as the moment of inertia

$$\Theta = m_0 r_0^2 \tag{45.13}$$

should be introduced as parameters. The constant γ expressed in these quantities is thus of the form

$$\gamma = \frac{4\pi^2 \nu \Theta}{h} . \tag{45.14}$$

143

Inserting this in (45.11) the following series is obtained for the energy E:

$$E = -\frac{C^2}{4D} + hv\left(n_r + \frac{1}{2}\right) + \frac{h^2}{8\pi^2\Theta}\left(l + \frac{1}{2}\right)^2 - \frac{3h^2}{8\pi^2\Theta}\left(n_r + \frac{1}{2}\right)^2$$

$$- \frac{3h^3}{32\pi^4v\Theta^2}\left(n_r + \frac{1}{2}\right)\left(l + \frac{1}{2}\right)^2 + \dots \tag{45.15}$$

The first term is the constant energy value corresponding to the classical limiting case, the second term equals the expression for the energy of the linear oscillator, and by the substitution of the $n_r = 0, 1, 2, 3 \dots$ values it yields the different vibrational energies, the third term yields the rotational energies corresponding to the orbital quantum numbers $l = 0, 1, 2, 3, \dots$ It can be proved that the fourth term originates from the anharmonicity of the oscillations, whereas the fifth, the last one written down in eq. (45.15), is produced by the coupling of the vibrational and rotational excitations.

The lowest energy corresponding to the ground state of the molecule is obtained for $n_r = l = 0$. In the case of the ground state the expansion (45.15) assumes the form

$$E_0 = -\frac{C^2}{4D} + \frac{hv}{2} - \frac{h^2}{16\pi^2\Theta} - \frac{3h^3}{256\pi^4v\Theta^2} + \dots$$

This expression yields the energy E_0 as a series in powers of the Planck constant h. The first term is the classical energy corresponding to $h = 0$, the second one is the zero-point vibrational energy corresponding to the frequency v [see the discussion following (32.22)], further terms describe the quantum-mechanical effects of higher order.

46. The Rigid Rotator

The rigid rotator is a mass point that can move around another point at a constant distance. Thus in this case the mass point moves on a spherical surface, in contrast with the plane rotator, where the mass point stays permanently in a plane and moves along a circular orbit.

The Schrödinger equation of the rigid rotator can be found easily. If the mass of the point is m_0 and its distance from the centre of rotation is r_0, i.e. $r = r_0$, then, as $V = 0$, the Schrödinger equation is

$$\nabla^2\psi + \frac{8\pi^2m_0}{h^2}E\psi = 0. \tag{46.1}$$

Taking into account that the equation does not depend on r, we find in polar coordinates

$$\frac{1}{r_0^2}\left[\frac{\partial^2\psi}{\partial\vartheta^2} + \cot\vartheta\,\frac{\partial\psi}{\partial\vartheta} + \frac{1}{\sin^2\vartheta}\,\frac{\partial^2\psi}{\partial\phi^2}\right] + \frac{8\pi^2 m_0}{h^2}E\psi = 0. \quad (46.2)$$

If the moment of inertia $m_0 r_0^2$ is denoted by Θ, it follows that

$$\frac{\partial^2\psi}{\partial\vartheta^2} + \cot\vartheta\,\frac{\partial\psi}{\partial\vartheta} + \frac{1}{\sin^2\vartheta}\,\frac{\partial^2\psi}{\partial\phi^2} + \frac{8\pi^2\Theta}{h^2}E\psi = 0. \quad (46.3)$$

This equation was already met under (37.7), and it was shown that its solution is

$$\psi = Y_l^m(\vartheta, \phi). \quad (46.4)$$

Furthermore, it was demonstrated that the coefficient of ψ in the last term on the left-hand side has to be equal to $l(l + 1)$, and thus

$$\frac{8\pi^2\Theta}{h^2}E = l(l + 1),$$

where l is an integer. Thus the energy eigenvalues are

$$E_l = l(l + 1)\frac{h^2}{8\pi^2\Theta}. \quad (46.5)$$

This result can be applied to diatomic molecules if Θ means the moment of inertia for an axis going through the centre of rotation, i.e. the centre of mass, and being perpendicular to the line joining the nuclei. As seen from the rotational spectra of the diatomic molecules the formula received for E_l is in agreement with the experimental findings, in contrast to the Bohr theory, where we found

$$E_l = l^2\frac{h^2}{8\pi^2\Theta}$$

(here we write l instead of the rotational quantum number m used in the Bohr theory).

It has to be emphasized that according to quantum mechanics the energy levels of the plane rotator differ from those of the rigid rotator. In the formula of the energy eigenvalue l^2 is found in case of the plane rotator, whereas $l(l + 1)$ stands for the rigid rotator.

47. The Two-body Problem

In the preceding sections the motion of a mass point m_0 in a given external potential-field was investigated. Now we are going to prove that these results can be applied to the description of systems consisting of two particles, where these particles are interacting.

If the masses of the two particles are m_1 and m_2, and their interaction potential depends only on the difference of their coordinates, $V = V(x_1 - x_2, y_1 - y_2, z_1 - z_2)$, then the energy expression of the two-body problem, containing the interaction and the kinetic energy term, may be written in the form

$$H = \frac{1}{2m_1}(p_{1x}^2 + p_{1y}^2 + p_{1z}^2) + \frac{1}{2m_2}(p_{2x}^2 + p_{2y}^2 + p_{2z}^2) + V(x_1 - x_2, y_1 - y_2, z_1 - z_2).$$

$$(47.1)$$

The Schrödinger equation can be constructed according to the general rule of (24.8). The eigenfunction ψ is a function of the coordinates x_1, y_1, z_1 and x_2, y_2, z_2. The Schrödinger equation is

$$-\frac{h^2}{8\pi^2 m_1}\left(\frac{\partial^2 \psi}{\partial x_1^2} + \frac{\partial^2 \psi}{\partial y_1^2} + \frac{\partial^2 \psi}{\partial z_1^2}\right) - \frac{h^2}{8\pi^2 m_2}\left(\frac{\partial^2 \psi}{\partial x_2^2} + \frac{\partial^2 \psi}{\partial y_2^2} + \frac{\partial^2 \psi}{\partial z_2^2}\right)$$

$$+ V(x_1 - x_2, y_1 - y_2, z_1 - z_2)\,\psi = E\psi. \qquad (47.2)$$

It is reasonable to introduce as new variables the coordinates of the centre of gravity of the two particles as well as the relative coordinates relating the first particle to the second one. The coordinates of the centre of gravity are

$$X = \frac{m_1 x_1 + m_2 x_2}{m_1 + m_2}, \qquad Y = \frac{m_1 y_1 + m_2 y_2}{m_1 + m_2}, \qquad Z = \frac{m_1 z_1 + m_2 z_2}{m_1 + m_2}, \quad (47.3)$$

whereas the relative coordinates are

$$x = x_1 - x_2, \quad y = y_1 - y_2, \quad z = z_1 - z_2. \qquad (47.4)$$

Applying the transformation rules of the differential operators

$$\frac{\partial}{\partial x_1} = \frac{m_1}{m_1 + m_2}\frac{\partial}{\partial X} + \frac{\partial}{\partial x}, \qquad \frac{\partial}{\partial x_2} = \frac{m_2}{m_1 + m_2}\frac{\partial}{\partial X} - \frac{\partial}{\partial x}, \quad (47.5)$$

(similar expressions hold for the y- and z-coordinates) the Schrödinger equation becomes

$$-\frac{h^2}{8\pi^2 M}\left(\frac{\partial^2\psi}{\partial X^2}+\frac{\partial^2\psi}{\partial Y^2}+\frac{\partial^2\psi}{\partial Z^2}\right)$$

$$-\frac{h^2}{8\pi^2 m}\left(\frac{\partial^2\psi}{\partial x^2}+\frac{\partial^2\psi}{\partial y^2}+\frac{\partial^2\psi}{\partial z^2}\right)+V(x,y,z)\,\psi=E\psi. \qquad (47.6)$$

Here $M=m_1+m_2$ is the total mass of the system, $m=(m_1 m_2)/(m_1+m_2)$ is the reduced mass of the two particles. The method of separation already used earlier can be applied to this new form of the Schrödinger equation. Let us ask for ψ in the following form:

$$\psi(X,Y,Z,x,y,z)=\chi(X,Y,Z)\,\phi(x,y,z).$$

After substitution and division by ψ;

$$\left[-\frac{h^2}{8\pi^2 M}\frac{1}{\chi}\left(\frac{\partial^2\chi}{\partial X^2}+\frac{\partial^2\chi}{\partial Y^2}+\frac{\partial^2\chi}{\partial Z^2}\right)\right]$$

$$+\left[-\frac{h^2}{8\pi^2 m}\frac{1}{\phi}\left(\frac{\partial^2\phi}{\partial x^2}+\frac{\partial^2\phi}{\partial y^2}+\frac{\partial^2\phi}{\partial z^2}\right)+V(x,y,z)\right]=E.$$

In the first square brackets only coordinates of the centre of mass, in the second ones only the relative coordinates are present. It is evident that this equation holds only if the terms in both square brackets are constant. Thus,

$$-\frac{h^2}{8\pi^2 M}\left(\frac{\partial^2\chi}{\partial X^2}+\frac{\partial^2\chi}{\partial Y^2}+\frac{\partial^2\chi}{\partial Z^2}\right)=\varepsilon_k\chi, \qquad (47.7)$$

$$-\frac{h^2}{8\pi^2 m}\left(\frac{\partial^2\phi}{\partial x^2}+\frac{\partial^2\phi}{\partial y^2}+\frac{\partial^2\phi}{\partial z^2}\right)+V(x,y,z)\,\phi=\varepsilon\phi, \qquad (47.8)$$

$$\varepsilon_k+\varepsilon=E. \qquad (47.9)$$

The first equation is the same as the Schrödinger equation of the free mass point (30.4), and so the solution is

$$\chi=e^{ik(\alpha X+\beta Y+\gamma Z)}, \qquad \varepsilon_k=\frac{h^2 k^2}{8\pi^2 M}. \qquad (47.10)$$

This means that the centre of mass moves absolutely freely, its state is described by a plane wave. According to (30.8) the momentum of the centre of mass is: $p=hk/2\pi$. If the centre of mass of the two-body system is regarded as stationary, then $p=0$, thus $k=0$ and according to (47.10) $\chi=1$

147

and $\varepsilon_k = 0$. Equation (47.8), describing the motion of the relative coordinates, is very interesting. This is, namely, equal to the form (24.3) of the Schrödinger equation describing the motion of a single particle in an external potential field V. The two-body problem has thus been reduced to the solution of the one-particle Schrödinger equation, the only difference is that the interaction potential acting between the two particles has to be written instead of potential V of the external field, and that the reduced mass of the two particles takes the place of mass m_0.

The theory of the H-atom has been treated in Section 44 by regarding the nucleus fixed in the centre of the coordinate system. Now it is easy to take into consideration relations due to the motion of the nucleus. According to (44.18) the energy levels of the H-atom are

$$E_n = - \frac{2\pi^2 m_0 Z^2 e^4}{h^2 n^2} .$$

Substituting the reduced mass of the electron–proton system in place of m_0,

$$m = \frac{m_0 M_0}{m_0 + M_0} = \frac{m_0}{1 + m_0/M_0}$$

the exact values of the energy levels of the H-atom are obtained. The only difference is that the E_n values have to be multiplied by

$$\frac{1}{1 + m_0/M_0} = 0.999\ 4556 .$$

This result agrees with (8.16) in every respect.

The forces keeping the atoms of a diatomic molecule together can be regarded, in good approximation, as harmonic, and can be described by the following potential:

$$V = \frac{1}{2} k(r - r_0)^2 = \frac{1}{2} kx^2 .$$

Here r is the distance between the two atoms, r_0 is the equilibrium distance between the nuclei, x is a measure of the displacement from the equilibrium position. This potential is the same as that of the harmonic oscillator discussed in Section 32. This means that also the diatomic molecule can be regarded as a harmonic oscillator, the energy levels of the oscillator correspond with the vibrational excitations of the molecule. These levels can be calculated according to (32.22) and (32.1) as follows:

$$E_n = h\nu \left(n + \frac{1}{2}\right) = \frac{h}{2\pi} \sqrt{\frac{k}{m}} \left(n + \frac{1}{2}\right),$$

where m is replaced by the reduced mass of the two atoms. From the investigation of the vibrational spectrum the frequency v can be determined, and from this one might reach conclusions about the force holding the molecules together. Thus, for example in the case of the HCl molecule $hv = $ $=0.358$ eV is obtained from vibrational spectroscopy, the reduced mass of the molecule is $17/18$ of the proton mass. From this the cohesive force is

$$k = 4\pi^2 m v^2 = 4.81 \times 10^5 \text{ g s}^{-2},$$

corresponding to the constant of a rather strong macroscopic spring.

The rotational levels of a diatomic molecule can be determined from the energies of the rigid rotator (46.5). In the expression $\Theta = mr_0^2, r_0$ is the distance between the two nuclei, m is the reduced mass. We only mention that this expression of Θ equals the moment of inertia of the molecule, where the axis of the moment goes through the centre of mass of the two nuclei and is perpendicular to the line joining them. From the formulae

$$\Theta = m_1 r_1^2 + m_2 r_2^2,$$

$$r_1 = \frac{m_2}{m_1 + m_2} r_0, \qquad r_2 = \frac{m_1}{m_1 + m_2} r_0$$

it follows immediately that

$$\Theta = \frac{m_1 m_2}{m_1 + m_2} r_0^2 = m r_0^2 .$$

Expression (46.5) enables the experimental determination of the moment of inertia Θ by the investigation of the rotational levels. Knowing the reduced mass m the equilibrium distance between the nuclei r_0 can be determined. For instance, in case of the HCl molecule we find $r_0 = 1.29$ Å.

Finally we mention that in Section 42 the same method was applied. Namely, the problem of the deuteron can be traced back to the one-body Schrödinger equation written with an exponential potential, only the reduced mass of the neutron–proton system (eq. (42.7)) takes the place of the mass.

CHAPTER 5

Scattering Problems

48. Introduction to the Theory of Scattering. Cross-sections

In the preceding sections our attention was directed towards the bound states of a mass point and towards the calculation of the energy eigenvalues E. If, according to general practice, the zero point of the energy scale was placed at the value of the potential reached at infinity, then the energy values of the bound states were always negative. States of positive energy describe particles that move through the potential-field and during this suffer a scattering. Now, these states will be considered. In the case of scattering problems the energy spectrum is continuous: each positive energy value is an eigenvalue of the Schrödinger equation, the energy of the bombarding particle is arbitrary, its value is determined by the initial conditions of the experiment.

In the following, not the investigation of the E energy-value, but that of the ψ wave function will be our task. The basic task of the scattering problem is the following. Particles with an energy E, forming a parallel beam with a particle-current density s_0, are incident onto a potential field described by the potential $V(x, y, z)$. It is supposed that this potential differs from zero significantly only in a finite part of space.* The question is what is the number of the particles scattered per unit time into a solid angle $d\Omega$ characterized by the polar angles ϑ and ϕ counted from the direction of the incident beam.

The number of particles scattered in the solid angle $d\Omega$ is proportional to the current density s_0 and the dimensions of the solid angle:

$$dn = \sigma(\vartheta, \phi) s_0 d\Omega. \tag{48.1}$$

The constant of proportionality $\sigma(\vartheta, \phi)$ is called the *differential scattering cross-section*. This is, of course a function of ϑ and ϕ determining the direction of the scattered particles. The total number of scattered particles can be determined by integrating (48.1):

$$n = s_0 \int \sigma(\vartheta, \phi) \, d\Omega = \sigma s_0. \tag{48.2}$$

---

* Among the important fields only the Coulomb forces do not satisfy these conditions, therefore the theory of the Coulomb scattering will be discussed separately.

σ is the *total scattering cross-section*, it is the integral of the differential scattering cross-section taken over the total solid angle.

The basic task of the quantum-mechanical calculus of scattering is the determination of the scattering cross-section by means of the Schrödinger equation.

The Schrödinger equation of a particle moving in a field described by the potential V is

$$\nabla^2 \psi + \frac{8\pi^2 m_0}{h^2} \left(E - V(x, y, z) \right) \psi = 0 \,.$$

E is now a given positive energy value, equal to the initial kinetic energy of the bombarding particles. Introducing the notation

$$k = \sqrt{\frac{8\pi^2 m_0}{h^2} E} \quad \text{and} \quad U(x, y, z) = \frac{8\pi^2 m_0}{h^2} V(x, y, z) \qquad (48.3)$$

the Schrödinger equation becomes

$$\nabla^2 \psi + \left(k^2 - U(x, y, z) \right) \psi = 0 \,. \qquad (48.4)$$

It is obvious that the function ψ can be determined from this differential equation unambiguously only if the boundary conditions are also fixed. In case of scattering problems such solutions are required that contain at a large distance from the scattering centre plane waves describing the incident beam, and spherical waves describing the scattered particles. Let us fix the z-axis of the coordinate system in the direction of the incident beam of particles. In this case the boundary condition can be described by the following asymptotic relation:

$$\psi(x, y, z) \approx N\left[e^{ikz} + f(\vartheta, \phi)\, \frac{e^{ikr}}{r} \right], \quad \text{if } r \to \infty \,. \qquad (48.5)$$

The first term in the square brackets describes a plane wave advancing in the z direction. The momentum of the particles in this wave is $p = hk/2\pi$. The second term is a spherical wave, the amplitude f of which is direction-dependent. The normalizing constant N can be chosen so that the current density of the incident plane wave should be s_0. From this we find that

$$N = \sqrt{\frac{2\pi m_0 s_0}{hk}} \,.$$

The radial current density of the scattered ray can be determined from formula (27.9). We find that

$$s = \frac{h}{4\pi i m_0} N^2 |f(\vartheta, \phi)|^2 \left(\frac{e^{-ikr}}{r} \frac{\partial}{\partial r} \frac{e^{ikr}}{r} - \frac{e^{ikr}}{r} \frac{\partial}{\partial r} \frac{e^{-ikr}}{r} \right) = \frac{s_0}{r^2} |f(\vartheta, \phi)|^2.$$

The surface element dF being at a distance r from the origin is seen under a solid angle $d\Omega = dF/r^2$, the number of the scattered particles traversing through dF is

$$dn = s\,dF = sr^2\,d\Omega = |f(\vartheta, \phi)|^2 s_0\,d\Omega.$$

Comparing this with (48.1) we see that the differential cross-section of scattering is

$$\sigma(\vartheta, \phi) = |f(\vartheta, \phi)|^2. \qquad (48.6)$$

Thus if we succeed in determining the solution of the wave equation (48.4) showing an asymptotic behaviour like (48.5), then the cross-section $\sigma(\vartheta, \phi)$ can be obtained immediately from the amplitude of the spherical wave $f(\vartheta, \phi)$. The problem thus reduces to the determination of the amplitude $f(\vartheta, \phi)$.

49. The Partial-wave Method

In the following only the problem of the scattering by a central potential field will be discussed where the potential V depends only on the distance r from the centre of force. In this case it seems reasonable to use the so-called *partial-wave method*, by which the wave function (48.4) can be reduced to the solution of a normal single-variable differential equation. For this reason let us expand the function ψ in terms of the angular-momentum eigenfunctions $Y_l^m(\vartheta, \phi)$. As both the incident beam and the scattering potential field are symmetrical around the z-axis, ψ is independent of the azimuth angle ϕ. This implies that in the series expansion only functions Y_l^m independent of ϕ will be present; these belong to the magnetic quantum number $m = 0$, and apart from a constant factor, are equal to the Legendre polynomials $P_l(\cos \vartheta)$. Thus the series expansion is

$$\psi = \frac{1}{r} \sum_{l=0}^{\infty} C_l u_l(r) P_l(\cos \vartheta). \qquad (49.1)$$

Substituting this in the wave equation (48.4) we obtain the following differential equation for the function $u_l(r)$:

$$\frac{d^2 u_l(r)}{dr^2} + \left[k^2 - U(r) - \frac{l(l+1)}{r^2} \right] u_l(r) = 0. \qquad (49.2)$$

153

The solution of this second-degree differential equation becomes unambiguous if two initial conditions are set. In order to determine these let us investigate the behaviour of $u_l(r)$ in the immediate vicinity of the centre of the potential field. If $r \to 0$, then in the square brackets of (49.2) the k^2 and $U(r)$ terms will be small in comparison with $l(l + 1)/r^2$, this means that the differential equation becomes asymptotically:

$$\frac{d^2 u_l(r)}{dr^2} = \frac{l(l + 1)}{r^2} u_l(r) , \qquad \text{if } r \to 0 . \qquad (49.3)$$

It is simple to solve this equation; let us try the solution in the form of $u_l(r) = r^p$. After substitution the equation $p(p - 1) = l(l + 1)$ is found, and from this two solutions are gained for p: $p_1 = l + 1$ and $p_2 = -l$. Hence the general solution of (49.3) is:

$$u_l(r) \approx K_1 r^{l+1} + K_2 \frac{1}{r^l} , \qquad \text{if } r \to 0 . \qquad (49.4)$$

The wave function ψ has to be everywhere bounded; it complies, however, with this requirement at $r = 0$ only if $u_l(0) = 0$. Therefore in (49.4) $K_2 = 0$. K_1 can thus be regarded as a normalizing constant in $u_l(r)$, and can be taken $K_1 = +1$, arbitrarily. According to this the solution of (49.2) is made unambiguous by the following boundary conditions:

$$u_l(r) \approx r^{l+1}, \qquad \text{as } r \to 0 . \qquad (49.5)$$

In the following this solution $u_l(r)$ will be supposed to be known. The concrete determination of $u_l(r)$ is, of course, performed by solving the differential equation (49.2); this is possible, disregarding the simplest potential functions, only by the method of numerical integration.

Knowing the $u_l(r)$ functions the wave function ψ of the scattering problem can be built up according to (49.1). The C_l coefficients of the expansion have to be determined in such a way that ψ should fulfil the boundary conditions of (48.5). For determining the C_l first we have to investigate the behaviour of $u_l(r)$ for very large values of r. Very far from the scattering centre, according to our supposition, the potential function $U(r)$ can be regarded as zero, and also $l(l + 1)/r^2$ becomes negligible. For large r the form of (49.2) thus becomes

$$\frac{d^2 u_l(r)}{dr^2} + k^2 u_l(r) = 0 , \qquad \text{as } r \to \infty . \qquad (49.6)$$

The general solution obviously is

$$u_l(r) \approx \frac{1}{k} A_l \sin\left(kr - \frac{l\pi}{2} + \delta_l\right), \qquad \text{as } r \to \infty . \tag{49.7}$$

The integration constants

$$A_i = A_l(k) \quad \text{and} \quad \delta_l = \delta_l(k)$$

cannot be chosen arbitrarily, as the equations (49.2) and (49.5) determine them together with the function $u_l(r)$ and from comparing the (numerical) solution with (49.7) A_l and δ_l can be simply determined.

Thus in the following A_l and δ_l can be taken to be known quantities. If (49.7) is written into the series expansion (49.1) of the total wave function, then we obtain for ψ the following asymptotic behaviour:

$$\psi \approx \frac{1}{kr} \sum_{l=0}^{\infty} C_l A_l \left[\cos \delta_l \sin\left(kr - \frac{l\pi}{2}\right) + \sin \delta_l \cos\left(kr - \frac{l\pi}{2}\right)\right] P_l(\cos \vartheta),$$

$$\text{as } r \to \infty . \tag{49.8}$$

To compare this with expression (48.5), the latter has to be expanded in a series of Legendre polynomials $P_l(\cos \vartheta)$. The series expansion of the plane wave is taken from the theory of the Bessel functions:

$$e^{ikz} = e^{ikr \cos \vartheta} = \sum_{l=0}^{\infty} (2l + 1) i^l \left(\frac{\pi}{2kr}\right)^{\frac{1}{2}} J_{l+\frac{1}{2}}(kr) P_l(\cos \vartheta) .$$

The asymptotic form of the Bessel function $J_{l+\frac{1}{2}}$, as $r \to \infty$ is also well known:

$$J_{l+\frac{1}{2}}(kr) \approx \left(\frac{2}{\pi kr}\right)^{\frac{1}{2}} \sin\left(kr - \frac{l\pi}{2}\right), \qquad \text{as } r \to \infty .$$

Thus for the series expansion of the plane wave the following asymptotic expression is valid:

$$e^{ikz} \approx \frac{1}{kr} \sum_{l=0}^{\infty} (2l + 1) i^l \sin\left(kr - \frac{l\pi}{2}\right) P_l(\cos \vartheta) . \tag{49.9}$$

In the case of a central potential the amplitude f of the scattered spherical wave will be independent of the azimuth angle ϕ; it will be a function only of the polar angle ϑ. We expand also $f(\vartheta)$ in a series of the Legendre poly-

155

nomials $P_l(\cos \vartheta)$:

$$f(\vartheta) = \sum_{l=0}^{\infty} B_l P_l(\cos \vartheta).\tag{49.10}$$

Now it is simple to write the required series expansion for (48.5). According to (49.9) and (49.10) we find that

$$\psi \approx N \sum_{l=0}^{\infty} \left[(2l + 1)i^l \frac{\sin\left(kr - \frac{l\pi}{2}\right)}{kr} + B_l \frac{e^{ikr}}{r} \right] P_l(\cos \vartheta) =$$

$$= \frac{N}{kr} \sum_{l=0}^{\infty} i^l \left[(2l + 1 + ikB_l) \sin\left(kr - \frac{l\pi}{2}\right) + kB_l \cos\left(kr - \frac{l\pi}{2}\right) \right] P_l(\cos \vartheta).\tag{49.11}$$

The asymptotic behaviour of (49.8) will agree with the expression (49.11) prescribed as a boundary condition if in both expansions the coefficients of $\sin (kr - l\pi/2)$ and $\cos (kr - l\pi/2)$ agree:

$$C_l A_l \cos \delta_l = N i^l (2l + 1 + ikB_l),$$

$$C_l A_l \sin \delta_l = N i^l k B_l.\tag{49.12}$$

As already mentioned, eqs. (49.2) and (49.5) define the quantities A_l and δ_l. In the two equations of (49.12) only the coefficients C_l and B_l of the series expansion can be regarded as unkown. The solution of the system of equations is

$$C_l = \frac{(2l + 1) N i^l}{A_l} e^{i\delta_l},$$

$$B_l = \frac{2l + 1}{2ik} (e^{i\delta_l} - 1).\tag{49.13}$$

We are interested primarily in B_l, the coefficient of the series expansion of the scattering amplitude $f(\vartheta)$. Writing the above form of B_l in (49.10), it yields

$$f(\vartheta) = \frac{1}{2ik} \sum_{l=0}^{\infty} (2l + 1) (e^{2i\delta_l} - 1) P_l(\cos \vartheta).\tag{49.14}$$

This formula means the solution of the scattering problem. We succeeded in expressing the scattering amplitude $f(\vartheta)$ in terms of the phase-constants δ_l of the $u_l(r)$ functions obtained as the solution of the ordinary differential equation (49.2). According to (48.6) the differential cross-section of scattering is

$$\sigma(\vartheta) = \frac{1}{4k^2} \left| \sum_{l=0}^{\infty} (2l + 1)\, (e^{2i\delta_l} - 1)\, P_l(\cos \vartheta) \right|^2 . \tag{49.15}$$

The total cross-section is gained by integrating this over ϑ. Using the integrals

$$\int_0^\pi P_{l'}(\cos \vartheta)\, P_l(\cos \vartheta) \sin \vartheta\, d\vartheta = 0 , \qquad \text{if } (l \neq l') ,$$

$$\int_0^\pi P_l^2(\cos \vartheta) \sin \vartheta\, d\vartheta = \frac{2}{2l + 1}$$

known from the theory of the Legendre polynomials we find

$$\sigma = \int_0^\pi \sigma(\vartheta)\, 2\pi \sin \vartheta\, d\vartheta = \frac{4\pi}{k^2} \sum_{l=0}^{\infty} (2l + 1) \sin^2 \delta_l . \tag{49.16}$$

The equations (49.15) and (49.16) for the cross-sections are specially adequate for concrete calculations in cases when the summation for l converges rapidly, when it is enough to restrict the calculations to the first few l values. This is the case if the energy of the bombarding particles is low. It has to be mentioned that the terminology introduced for the classification of the atomic terms is usually used in scattering calculus as well. The terms in (49.14) and (49.16) corresponding to $l = 0, 1, 2, 3, \ldots$ are called the $s, p, d, f, \ldots$ *scattering amplitude*, and *cross-section*, respectively.

The total cross-section shows a very interesting relation to the value of the scattering amplitude $f(\vartheta)$ at $\vartheta = 0$. For this one obtains from (49.14), using the fact that $P_l(1) = 1$:

$$f(0) = \frac{1}{2ik} \sum_{l=0}^{\infty} (2l + 1)\, (e^{2i\delta_l} - 1) = \frac{1}{k} \sum_{l=0}^{\infty} (2l + 1) \sin \delta_l (\cos \delta_l + i \sin \delta_l) .$$

The total cross-section is, according to (49.16), $4\pi/k$ times the imaginary part of this expression

$$\sigma = \frac{4\pi}{k} \operatorname{Im} f(0) . \tag{49.17}$$

Usually this important relation is called the "optical theorem". Its physical content is the following: from the incident particle current $s_0 n = \sigma s_0$ is scattered into the different directions of space; behind the scattering centre (thus in direction $\vartheta = 0$) the intensity of the beam is weakened to this extent. This drop in intensity is described by the interference of the two terms of the asymptotic expression (48.5), the plane and the spherical waves. This interference term is proportional to the amplitude $f(0)$ of the forward scattering, and this has to supply the drop in intensity expressed by the cross-section σ. This relation between $f(0)$ and σ is expressed by the optical theorem.

50. Scattering by a Three-dimensional Potential Well

As a simple example let us investigate the problem of scattering by a three-dimensional potential well. The potential is given by expression (40.1), where a is the radius of the well and V_0 is its depth. The differential equation (49.2) of the partial waves $u_l(r)$ shows now the following form:

$$\frac{d^2u_l}{dr^2} + \left(\alpha^2 - \frac{l(l+1)}{r^2}\right)u_l = 0, \quad \text{if } r < a,$$

$$\frac{d^2u_l}{dr^2} + \left(k^2 - \frac{l(l+1)}{r^2}\right)u_l = 0, \quad \text{if } r > a,$$

(50.1)

where

$$\alpha^2 = \frac{8\pi^2 m_0}{h^2}(E + V_0) = k^2 + \frac{8\pi^2 m_0}{h^2}V_0. \tag{50.2}$$

Equations (50.1) are in close connection with the differential equation of the Bessel functions of order $l + \frac{1}{2}$. Two linearly independent solutions of the second equation, that is for $r > a$, are

$$j_l(kr) = \left(\frac{\pi kr}{2}\right)^{\frac{1}{2}} J_{l+\frac{1}{2}}(kr) \quad \text{and} \quad n_l(kr) = \left(\frac{\pi kr}{2}\right)^{\frac{1}{2}} N_{l+\frac{1}{2}}(kr). \tag{50.3}$$

(Here $J_{l+\frac{1}{2}}$ and $N_{l+\frac{1}{2}}$ denote the Bessel functions of first and second order, respectively.) The linear combination of the functions (50.3) will give the general solution

$$u_l(r) = b_1 j_l(kr) + c_1 n_l(kr), \quad \text{if } r > a. \tag{50.4}$$

The solution for $r < a$ can be written in a similar form. The only difference is that k is replaced by α.

$$u_l(r) = b_2 j_l(\alpha r) + c_2 n_l(\alpha r), \quad \text{if } r < a. \tag{50.5}$$

The constants of integration b_1, c_1 and b_2, c_2 are determined by the boundary conditions. Let us first satisfy the initial condition (49.5). For very small arguments the following asymptotic relations hold for the Bessel functions:*

$$J_{l+\frac{1}{2}}(x) \approx \left(\frac{2}{\pi}\right)^{\frac{1}{2}} \frac{x^{l+\frac{1}{2}}}{(2l+1)!!}$$

$$\text{if } x \to 0 \qquad (50.6)$$

$$N_{l+\frac{1}{2}}(x) \approx -\left(\frac{2}{\pi}\right)^{\frac{1}{2}} \frac{(2l-1)!!}{x^{l+\frac{1}{2}}} .$$

The asymptotic behaviour of the solution (50.5) is

$$u_l(r) \approx b_2 \frac{(\alpha r)^{l+1}}{(2l+1)!!} - c_2 \frac{(2l-1)!!}{(\alpha r)^l} .$$

Comparing this with the initial condition (49.5) we obtain the following values for b_2 and c_2:

$$b_2 = \frac{(2l+1)!!}{\alpha^{l+1}} \quad \text{and} \quad c_2 = 0 . \qquad (50.7)$$

The constants b_1 and c_1 are determined by the requirements that both u_l and du_l/dr should be continuous at $r = a$:

$$\frac{(2l+1)!!}{\alpha^{l+1}} j_l(\alpha a) = b_1 j_l(ka) + c_1 n_l(ka) ,$$

$$\qquad (50.8)$$

$$\frac{(2l+1)!!}{k\alpha^l} j_l'(\alpha a) = b_1 j_l'(ka) + c_1 n_l'(ka) .$$

The solution of the system of equations is

$$b_1 = -\frac{(2l+1)!!}{\alpha^{l+1}k} \cdot \frac{\alpha j_l'(\alpha a) n_l(ka) - k j_l(\alpha a) n_l'(ka)}{j_l(ka) n_l'(ka) - j_l'(ka) n_l(ka)} ,$$

$$\qquad (50.9)$$

$$c_1 = +\frac{(2l+1)!!}{\alpha^{l+1}k} \cdot \frac{\alpha j_l'(\alpha a) j_l(ka) - k j_l(\alpha a) j_l'(ka)}{j_l(ka) n_l'(ka) - j_l'(ka) n_l(ka)} .$$

From this the partial waves $u_l(r)$ are fully determined. Let us now investigate how $u_l(r)$ behaves at large distances r and let us determine the phase

---

* According to general practice, the following product is denoted by the double factorial sign: $(2l+1)!! = 1 \cdot 3 \cdot 5 \ldots (2l+1)$.

constant δ_l. First the formula describing the asymptotic behaviour of the Bessel functions is recalled:

$$J_{l+\frac{1}{2}}(x) \approx \left(\frac{2}{\pi x}\right)^{\frac{1}{2}} \sin\left(x - \frac{l\pi}{2}\right)$$

$$\text{if } |x| \to \infty \quad (50.10)$$

$$N_{l+\frac{1}{2}}(x) \approx -\left(\frac{2}{\pi x}\right)^{\frac{1}{2}} \cos\left(x - \frac{l\pi}{2}\right),$$

Substituting (50.10) in (50.4) the following asymptotic expressions are obtained:

$$u_l(r) \approx b_1 \sin\left(kr - \frac{l\pi}{2}\right) - c_1 \cos\left(kr - \frac{l\pi}{2}\right)$$

$$= (b_1^2 + c_1^2)^{\frac{1}{2}} \sin\left(kr - \frac{l\pi}{2} - \arctan\frac{c_1}{b_1}\right), \quad \text{if } r \to \infty . \quad (50.11)$$

Comparing this with (49.7) we can state that

$$A_l = k(b_1^2 + c_1^2)^{\frac{1}{2}} \quad \text{and} \quad \tan \delta_l = -\frac{c_1}{b_1}. \quad (50.12)$$

By using the expressions (50.9) the final result is

$$\tan \delta_l = \frac{\alpha j_l'(\alpha a) j_l(ka) - k j_l(\alpha a) j_l'(ka)}{\alpha j_l'(\alpha a) n_l(ka) - k j_l(\alpha a) n_l'(ka)} . \quad (50.13)$$

In Fig. 23 the phase constant δ_l is shown as function of k for the partial waves $l = 0$, 1 and 2. The total cross-sections computed from this are seen in Fig. 24. For very low bombarding energies ($ka \ll 1$) the cross-section of the p-wave (and that of all the other waves belonging to higher orbital quantum numbers) is negligibly small compared to the cross-section of the s-wave. For very high bombarding energies ($ka \gg 1$) the cross-section of every partial wave tends to zero. In the range of moderate energies a very interesting phenomenon can be observed. In the case of the special potential well investigated in Figs. 23 and 24 the phase angle of the p-wave increases rapidly near $ka = 0.7$ and in the meantime passes the value of $3\pi/2$. Meanwhile the total cross-section corresponding to the p-wave

$$\sigma_1 = \frac{12\pi}{k^2} \sin^2 \delta_1$$

takes its maximum value, $12\pi/k^2$. This produces the sharp maximum seen in Fig. 24. This phenomenon is called *resonance scattering*. Such resonances may occur, of course, for any partial wave corresponding to any l. The k-values where the resonance may occur depend on the special values of the parameters of the potential well. The general theory of the scattering resonances will be discussed in the next section.

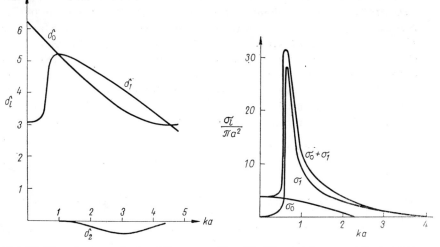

FIG. 23. The phase-constants of the s, p and d partial waves, as functions of k, for the case where

$$\frac{8\pi^2 m_0 a^2 V_0}{h^2} = 6.2$$

FIG. 24. The total cross-sections corresponding to the s and p partial waves, as functions of k, for the case where

$$\frac{8\pi^2 m_0 a^2 V_0}{h^2} = 6.2$$

51. Unified Theory of the Scattering and Bound States. Resonance Scattering

The bound quantum states occurring in a central potential-field have been considered in detail in Section 37. It was shown that the eigenvalues can be determined from the radial Schrödinger equation (37.17):

$$\frac{d^2 R}{dr^2} + \frac{2}{r}\frac{dR}{dr} - \frac{l(l+1)}{r^2}R + \frac{8\pi^2 m_0}{h^2}[E - V(r)]R = 0.$$

If the following abbreviations are introduced

$$\kappa = \sqrt{-\frac{8\pi^2 m_0}{h^2}E} \quad \text{and} \quad U(r) = \frac{8\pi^2 m_0}{h^2}V(r) \qquad (51.1)$$

161

and the substitution $R(r) = (1/r)u_l(r)$ is performed, then we obtain the differential equation:

$$\frac{d^2u_l(r)}{dr^2} + \left[-\kappa^2 - U(r) - \frac{l(l+1)}{r^2} \right] u_l(r) = 0 . \tag{51.2}$$

After the substitution $\kappa \to ik$ this form of the radial Schrödinger equation takes the form of the differential equation of the scattering problem (49.2). Furthermore, it can be stated that also the initial conditions for the function $u_l(r)$ are the same for the case of the scattering and the case of bound states. Namely, in both cases the asymptotic behaviour $u_l(r) = r^{l+1}$ for $r \to 0$ has to be prescribed, otherwise the wave function ψ would not stay finite at $r = 0$. Thus the problems of scattering and bound states are mathematically very similar, and this makes possible the common treatment of the two questions. The starting equation is the following differential equation

$$\frac{d^2u_l}{dr^2} + \left[q^2 - U(r) - \frac{l(l+1)}{r^2} \right] u_l = 0 , \tag{51.3}$$

where q is a complex number. The initial condition attached to this differential equation is

$$u_l(r) \approx r^{l+1}, \quad \text{if } r \to 0 . \tag{51.4}$$

Through this the solution of u_l becomes unambiguous. The solution satisfying (51.4) is called the "physical solution" of the differential equation (51.3), and is denoted by $u_l = u_l(q, r)$. Let us now investigate the asymptotic behaviour of u_l as $r \to \infty$. In this case in (51.3) $U(r)$ and $l(l+1)/r^2$ tend to zero rapidly, and the asymptotic form of the differential equation becomes

$$\frac{d^2u_l}{dr^2} + q^2 u_l = 0 . \tag{51.5}$$

It is well known that two special solutions of this equation are e^{iqr} and e^{-iqr}, and all the other solutions are linear combinations of these. The asymptotic behaviour of the physical solution is thus given by

$$u_l(q, r) \approx \frac{1}{2iq} \left[F_l(q) e^{iqr} + H_l(q) e^{-iqr} \right], \quad \text{if } r \to \infty \tag{51.6}$$

and for determining the coefficients $F_l(q)$ and $H_l(q)$ of the linear combination the differential equation (51.3) has to be solved (for instance by numerical methods) taking the initial condition (51.4) into consideration. If $U(r)$

is a concrete potential this can always be done, and in the following $F_l(q)$ and $H_l(q)$ will be supposed to be known.

Now some important characteristics of $F_l(q)$ and $H_l(q)$ will be discussed. Equations (51.3) and (51.4) remain unchanged if the substitution $q \to -q$ is performed; therefore also the physical solutions belonging to q and $-q$ will be the same:

$$u_l(q, r) = u_l(-q, r). \qquad (51.7)$$

This must apply, of course, also for $r \to \infty$, thus

$$\frac{1}{2iq} [F_l(q) e^{iqr} + H_l(q) e^{-iqr}] = - \frac{1}{2iq} [F_l(-q) e^{-iqr} + H_l(-q) e^{iqr}].$$

This equation is valid then, and only then, when

$$H_l(q) = - F_l(-q). \qquad (51.8)$$

This implies that it is not necessary to work with two coefficients, for if $F_l(q)$ is known (for all values of q) then, according to (51.8), $H_l(q)$ is known, too. In other words, the asymptotic behaviour of the physical solution is described by the formula

$$u_l(q, r) \approx \frac{1}{2iq} [F_l(q) e^{iqr} - F_l(-q) e^{-iqr}] \qquad (51.9)$$

which is more special in structure than (51.6).

Let us now form the conjugate complexes of the basic equations (51.3) and (51.4). As can be seen, $[u_l(q, r)]^*$ satisfies the differential equation and the initial condition for q^*. Thus

$$[u_l(q, r)]^* = u_l(q^*, r). \qquad (51.10)$$

From this it follows immediately that for real and for purely imaginary q the expression $u_l(q, r)$ will be real. (This rule is very significant for a concrete computation of u_l.) Using (51.10) the rule of conjugation of $F_l(q)$ can be determined as well: namely, (51.10) will be valid for $r \to \infty$ if, and only if

$$[F_l(q)]^* = F_l(-q^*). \qquad (51.11)$$

This rule is of great practical importance. It is enough to solve the differential equation (51.3) for the right-hand side of the complex q plane (Argand diagram), that is for $Re\ q \geq 0$, as on the left-hand side of the q-

plane $F_l(q)$ can be determined from (51.11). We put (51.11) for two special cases as well. If q is equal to a real value k, then the rule of conjugation is

$$[F_l(k)]^* = F_l(-k) .$$

and if q is a purely imaginary number, then

$$[F_l(i\kappa)]^* = F_l(i\kappa) .$$

That is, for purely imaginary q values, F_l will be a real number.

In the following the zero points of the function $F_l(q)$ will be of special importance. Let us denote one of the roots of the equation

$$F_l(q) = 0 \tag{51.12}$$

by q. According to rule (51.11) $F_l(-q^*) = [F_l(q)]^* = 0$: that is $-q^*$ is also a root. This means that the roots (51.12) lie in the Argand diagram symmetrically with respect to the imaginary axis. Even more can be stated for the roots lying in the lower half plane. If $q = k - i\kappa$ ($\kappa > 0$) is one of the roots lying in the lower half plane then according to (51.9) the asymptotic behaviour of the appropriate $u_l(q, r)$ function is

$$u_l(q, r) \approx -\frac{1}{2iq} F_l(-k + i\kappa) e^{-ikr} e^{-\kappa r}, \quad \text{if } r \to \infty . \tag{51.13}$$

Thus u_l vanishes exponentially if $r \to \infty$, as κ is a positive number. Multiplying the differential equation (51.3) by $[u_l(q, r)]^*$ and the conjugate complex of (51.3) by $u_l(q, r)$, and subtracting them from each other the result is

$$\frac{d}{dr}\left[u_l(q, r) \frac{d}{dr}[u_l(q, r)]^* - [u_l(q, r)]^* \frac{d}{dr} u_l(q, r) \right] = (q^2 - q^{*2}) \, | \, u_l(q, r) \, |^2 .$$

Integrating with respect to r from 0 to ∞:

$$\left[u_l(q, r) \frac{d}{dr}[u_l(q, r)]^* - [u_l(q, r)]^* \frac{d}{dr} u_l(q, r) \right]_0^\infty = (q^2 - q^{*2}) \int_0^\infty | \, u_l(q, r) \, |^2 \, dr .$$

The left-hand side vanishes as r^{2l+1} for $r \to 0$, because of (51.4). It vanishes as $e^{-2\kappa r}$ for $r \to \infty$, because of (51.13). At the same time the integral on the right-hand side is a non-zero finite number. Thus the equality can hold only if $q^2 = q^{*2}$, i.e. $q = \pm q^*$. On the other hand, only $q = -q^*$ is possible for $\kappa > 0$, thus $k = 0$ and $q = \pm i\kappa$.

Summarizing, the roots of equation $F_l(q)$ are distributed symmetrically with respect to the imaginary axis in the Im $q > 0$ half-plane and lie on the imaginary axis in the half-plane Im $q < 0$. Figure 25 gives a qualitative picture of the position of the roots.

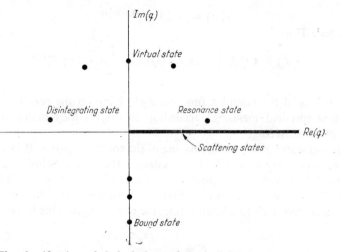

Fig. 25. The classification of the solutions of the radial Schrödinger equation. The dots denote the roots of the equation $F_l(q) = 0$ in the complex plane q

According to these the physical solutions $u_l(q, r)$ can be classified as follows:

1. Scattering state: q is a real, positive number: k. The corresponding function $u_l(k, r)$ will be equal to the solution of the scattering problem (49.2). Let us investigate the asymptotic behaviour of $u_l(k, r)$ for $r \to \infty$. The form of (51.9) for a real wave number k is:

$$u_l(k, r) \approx \frac{1}{2ik} \left[F_l(k) e^{ikr} - F_l(-k) e^{-ikr} \right]$$

$$= \frac{1}{2ik} \left[F_l(k) e^{ikr} - F_l^*(k) e^{-ikr} \right], \quad \text{as } r \to \infty. \quad (51.14)$$

Let us find the connection between the coefficient $F_l(k)$ determining the asymptotic behaviour and the phase-constant δ_l for scattering. For this reason formula (49.7) defining the phase constant is transcribed by the help of the Euler relation:

$$u_l(k, r) \approx \frac{1}{2ik} \left[A_l e^{i(\delta_l - l\pi/2)} e^{ikr} - A_l e^{-i(\delta_l - l\pi/2)} e^{-ikr} \right].$$ (51.15)

By comparing (51.14) and (51.15)

$$F_l(k) = A_l(k) \, e^{i(\delta_l(k) - l\pi/2)}$$ (51.16)

is reached. Thus,

$$A_l(k) = |F_l(k)| \quad \text{and} \quad \delta_l(k) = \frac{l\pi}{2} + \arg F_l(k).$$ (51.17)

The values of the complex function $F_l(k)$ lying on the real axis have an important physical meaning: according to (51.17) they define the phase constant of scattering.

2. Bound state: $q = -i\kappa$ equals one of the roots of $F_l(q) = 0$ lying on the negative imaginary axis. $u_l(-i\kappa, r)$ satisfies the radial Schrödinger equation of (51.2) as well as the conditions of being bounded: it vanishes as r^{l+1} for $r = 0$, and for $r \to \infty$ as $e^{-\kappa r}$. Such a solution of the radial Schrödinger equation corresponds to a bound state, the energy eigenvalue is from (51.1)

$$E = -\frac{h^2 \kappa^2}{8\pi^2 m_0}.$$ (51.18)

And so the roots of $F_l(q) = 0$ lying on the negative imaginary axis deliver the energy eigenvalues of the bound states.

3. Virtual state: $q = \pm k_0 + i\kappa$ is a root of the equation $F_l(q) = 0$ lying in the upper half of the complex q-plane. Those virtual states have a clear descriptive meaning when q lies very near the real axis; in the following we are concerned only with these. Let us first investigate a virtual state lying in the second quarter of the Argand diagram: $q = -k_0 + i\kappa$, where k_0 and κ are positive numbers. Expression (51.9) defining the asymptotic behaviour of u_l takes now, due to $F_l(q) = 0$, the following form:

$$u_l(q, r) \approx -\frac{1}{2iq} F_l(-q) e^{-iqr} = -\frac{1}{2iq} F_l(-q) e^{ik_0 r} e^{\kappa r} = \text{const.} \cdot e^{\kappa r} e^{ik_0 r}.$$ (51.19)

The $e^{ik_0 r}$ term describes an outgoing spherical wave the amplitude of which is modulated by $e^{\kappa r}$. (Of course, it is correct to speak about a modulated wave only if $e^{-ik_0 r}$ oscillates much more rapidly than $e^{\kappa r}$ changes, i.e. if $k_0 \gg \kappa$. This condition is, however, really satisfied if the root $q = -k_0 + i\kappa$

is near the real axis.) Let us construct the total state function corresponding to (51.19). For this reason the radial wave function $(1/r) u_l$ has to be augmented by the angle-dependent term $Y_l^m(\vartheta, \phi)$, and by the time-dependent term $e^{-(2\pi i/h)\varepsilon t}$ where the energy parameter ε is now $\varepsilon = h^2 q^2/8\pi^2 m_0$. The total wave function thus becomes

$$\Psi(x, y, z, t) = \frac{1}{r} u_l(q, r) Y_l^m(\vartheta, \phi) e^{-(2\pi i/h)\varepsilon t}$$

$$= \frac{1}{r} u_l(q, r) Y_l^m(\vartheta, \phi) e^{-(ih/4\pi m_0)(k_0^2-\kappa^2-2ik_0\kappa)t} .$$

Very far from the origin the asymptotic behaviour of the wave function is defined by (51.19):

$$\Psi(x, y, z, t) \approx -\frac{1}{2iqr} F_l(-q) e^{[\kappa r-(hk_0\kappa/2\pi m_0)t]} e^{i[k_0 r-(h/4\pi m_0)(k_0^2-\kappa^2)t]} Y_l^m(\vartheta, \phi).$$

$$(51.20)$$

In accordance with the discussion in Section 27 the corresponding particle density and radial current density is

$$\rho(x, y, z, t) = \Psi^*\Psi \approx \left| \frac{F_l(-q)}{2q} \right|^2 \cdot \frac{e^{2[\kappa r-(hk_0\kappa/2\pi m_0)t]}}{r^2} |Y_l^m|^2,$$

$$s_r(x, y, z, t) = \frac{h}{4\pi i m_0} \left(\Psi^* \frac{\partial \Psi}{\partial r} - \Psi \frac{\partial \Psi^*}{\partial r} \right)$$

$$\approx \frac{hk_0}{2\pi m_0} \left| \frac{F_l(-q)}{2q} \right|^2 \frac{e^{2[\kappa r-(hk_0\kappa/2\pi m_0)t]}}{r^2} |Y_l^m|^2 .$$

The most important physical characteristics of the state ψ can be read from these equations. Our expressions describe the field of particles leaving the origin. According to the relation $s_r = v\rho$ the radial velocity of the particles is

$$v = \frac{s_r}{\rho} = \frac{hk_0}{2\pi m_0} .$$

$$(51.21)$$

As k_0 is positive v will be positive, too; that is, the particles move away from the origin in a radial direction. The kinetic energy of the radial movement is

$$E_0 = \frac{1}{2} m_0 v^2 = \frac{h^2 k_0^2}{8\pi^2 m_0} .$$

$$(51.22)$$

The total intensity of the particle current crossing a spherical surface of radius r is

$$I(r, t) = \int s_r(x, y, z, t) \, r^2 \sin \vartheta \, d\vartheta \, d\phi = I_0 e^{-(2\pi/h)\Gamma(t-r/v)}, \qquad (51.23)$$

where I_0 is a constant independent of space and time, and

$$\Gamma = \frac{h^2 k_0 \kappa}{2\pi^2 m_0}. \qquad (51.24)$$

Thus according to our results ψ describes a field of particles flying away from the origin with a velocity v, while the intensity I of the source decreases exponentially with the time. Such a state is called a radioactively disintegrating state. The disintegration half-life, i.e. the time-interval T within which the intensity I is halved, is calculated from $e^{-(2\pi/h)\Gamma T} = \frac{1}{2}$

$$T = \frac{h \ln 2}{2\pi \Gamma} = \frac{\pi \ln 2 \cdot m_0}{h k_0 \kappa}. \qquad (51.25)$$

Let us now turn towards those virtual states that lie in the first quadrant of the q complex plane. Here $q = k_0 + i\kappa$, where k_0 and κ are positive numbers. Being a virtual state q satisfies the equation $F_l(q) = 0$. We investigate the number $F_l'(q)$, where the prime denotes a derivative. The absolute value of this should be denoted by B_l and the argument by $(\Phi_l - l\pi/2)$:

$$F_l'(q) = B_l e^{i(\Phi_l - l\pi/2)}. \qquad (51.26)$$

We suppose that the root q lies close to the real axis:

$$\kappa \ll k_0.$$

We consider the immediate surroundings of k_0 on the real axis (see Fig. 26) and determine on this section the phase-constant of scattering $(\delta_l(k))$. First we expand F_l in a Taylor series in the neighbourhood of the root q:

$$F_l(k) = F_l(q) + F_l'(q) \, (k - q) + \dots$$

As k and q lie close to each other, the higher powers of $(k - q)$ may be neglected. Because of (51.12) and (51.26), the expression for $F_l(k)$ then

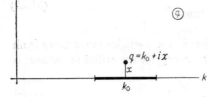

FIG. 26. The region of resonance scattering

becomes

$$F_l(k) = B_l e^{i(\Phi_l - l\pi/2)} (k - k_0 - i\kappa).$$ (51.27)

The scattering phase constant belonging to the wave number δ_l can be determined from $F_l(k)$ by the help of (51.17):

$$\delta_l(k) = \Phi_l - \arctan \frac{\kappa}{k - k_0}.$$ (51.28)

Thus the phase constant consists of two terms: the constant Φ_l is independent of k, the other term depends on k considerably:

$$\delta_l^{\text{res}}(k) = - \arctan \frac{\kappa}{k - k_0}.$$ (51.29)

This is the resonance term, its resonance value is at $k = k_0$ equal to $\pi/2$. The cross-section of resonance scattering is

$$\sigma_l^{\text{res}} = \frac{4\pi(2l + 1)}{k^2} \sin^2 \delta_l \simeq \frac{4\pi(2l + 1)}{k_0^2} \cdot \frac{\kappa^2}{(k - k_0)^2 + \kappa^2}.$$ (51.30)

In the first term the approximation $k \sim k_0$ is permissible. Let us express the cross-section by means of the energy of the bombarding particles $E = \hbar^2 k^2/8\pi^2 m_0$. If the notation (51.22) and (51.24) are introduced, (51.30) can be brought to the following form:

$$\sigma_l^{\text{res}} \simeq \frac{\text{const.}}{(E - E_0)^2 + \Gamma^2}.$$ (51.31)

The constant coefficient denotes a quantity independent of E. During the remodelling of the expression the terms with higher powers of $(E - E_0)$ have been neglected. Expression (51.31) is called the Breit–Wigner formula. The resonance cross-section corresponding to it is seen in Fig. 27. The resonance scattering shows a sharp maximum at $E = E_0$; the width of the resonance equals Γ.

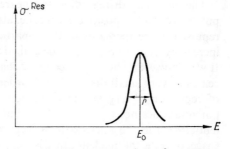

FIG. 27. The cross-section of resonance scattering. E_0 is the resonance energy, and S the width of the resonance

169

It has been shown that the roots of (51.12) lie symmetrically with respect to the imaginary axis. This implies that the disintegrating state corresponds to a scattering resonance, and vice versa. The kinetic energy E_0 of the particles flying out of the disintegrating state is just equal to the resonance energy of the scattering; the half-lifetime T of the disintegrating state has a simple relation with the half-width of the resonance:

$$T = \frac{h \ln 2}{2\pi \Gamma}.$$

By observing the disintegrating states we can reach conclusions about the position and width of the resonance and conversely: if a resonance maximum occurs in the scattering, then a corresponding disintegrating state always exists.

52. Radioactive α-Disintegration

During the α-disintegration helium nuclei, or α-particles, are emitted by the radioactive nucleid. Natural α-disintegration is shown, apart from an isotope of samarium, only by the heaviest elements. The limit is there very sharp; the heaviest stable element is the Bi isotope with a nuclear mass of 209, the lightest α-emitting nucleus is the Po with a nuclear mass of 210. The kinetic energy and the half-life of the different α-emitting elements are generally different. A characteristic relation found by Geiger and Nuttall exists between the energy of the α-particles and the half-life of the disintegrating element. This relation states that the kinetic energy of the α-particles decreases with increasing half-life of the α-emitting isotope. The reasons for this relation have been offered only by wave mechanics, and were first deduced by Gamow. In the following the Gamow theory will be considered.

The nucleus — disregarding the nearest surrounding of the nucleus — repulses every other positive charge, i.e. also an α-particle. The strength of this repulsive force increases as the α-particle gets nearer to the nucleus. This increasing repulsive force is described by the Coulomb potential $2Ze^2/r$. It was, however, shown in Section 2 that the repulsive action of the nucleus ceases at very small distances, the order of which is 10^{-12} cm. Here instead of repulsion a very strong attraction is observed, which means that the potential does not increase further, but decreases rapidly, producing a potential well in the nucleus that is separated by a potential barrier from the space outside the nucleus; this is demonstrated in Fig. 28a. For simplicity the potential will be approximated by the rectangular potential barrier of Fig. 28b.

First we write down the radial Schrödinger equation for the s wave corresponding to $l = 0$:

$$\frac{d^2u_0}{dr^2} + q^2u_0 = 0, \qquad \text{if } r < a_1,$$

$$\frac{d^2u_0}{dr^2} - p^2u_0 = 0, \qquad \text{if } a_1 < r < a_2, \qquad (52.1)$$

$$\frac{d^2u_0}{dr^2} + q^2u_0 = 0, \qquad \text{if } r > a_2.$$

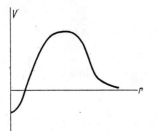

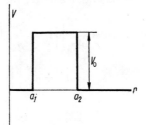

FIG. 28a. Potential well and barrier in the nucleus

FIG. 28b. Rectangular potential barrier

Here the notation

$$p = p(q) = \sqrt{\frac{8\pi^2 m_0}{h^2} V_0 - q^2} \qquad (52.2)$$

has been introduced. The general solution of the differential equations (52.1) follows immediately

$$u_0 = b_1 \sin(qr) + c_1 \cos(qr), \qquad \text{if } r < a_1,$$

$$u_0 = b_2 e^{p(r-a_1)} + c_2 e^{-p(r-a_1)}, \qquad \text{if } a_1 < r < a_2, \qquad (52.3)$$

$$u_0 = b_3 e^{iq(r-a_2)} + c_3 e^{-iq(r-a_2)}, \qquad \text{if } r > a_2.$$

Our first task is the determination of the constants of integration. From the initial condition (51.4) prescribed for the physical solution we reach

$$b_1 = \frac{1}{q} \quad \text{and} \quad c_1 = 0. \qquad (52.4)$$

The constants b_2 and c_2 are determined by the requirement that u_0 and its derivative should be continuous at $r = a_1$. Hence

$$b_2 = \frac{1}{2q} \left(\sin(qa_1) + \frac{q}{p} \cos(qa_1) \right),$$

$$c_2 = \frac{1}{2q} \left(\sin(qa_1) - \frac{q}{p} \cos(qa_1) \right). \tag{52.5}$$

The constants b_3 and c_3 are determined from the requirements of continuity at $r = a_2$. Introducing the distance $l = a_2 - a_1$, defining the thickness of the potential barrier, the result is as follows:

$$b_3 = \frac{1}{4q} \left[\left(1 - \frac{ip}{q}\right) \left(\sin(qa_1) + \frac{q}{p}\cos(qa_1)\right) e^{pl} \right.$$

$$\left. + \left(1 + \frac{ip}{q}\right) \left(\sin(qa_1) - \frac{q}{p}\cos(qa_1)\right) e^{-pl} \right],$$

$$c_3 = \frac{1}{4q} \left[\left(1 + \frac{ip}{q}\right) \left(\sin(qa_1) + \frac{q}{p}\cos(qa_1)\right) e^{pl} \right. \tag{52.6}$$

$$\left. + \left(1 - \frac{ip}{q}\right) \left(\sin(qa_1) - \frac{q}{p}\cos(qa_1)\right) e^{-pl} \right].$$

Let us now investigate the behaviour of u_0 as $r \to \infty$. The last expression of (52.3) is identical with the asymptotic formula of (51.9). Comparing these two expressions we may state that

$$\frac{1}{2iq} F_0(q) = b_3 e^{-iqa_2} \quad \text{and} \quad -\frac{1}{2iq} F_0(-q) = c_3 e^{iqa_2}.$$

By substituting (52.6), the following end result is obtained:

$$F_0(q) = \frac{i}{2} e^{-iqa_2} \left[\left(1 - \frac{ip}{q}\right) \left(\sin(qa_1) + \frac{q}{p}\cos(qa_1)\right) e^{pl} \right.$$

$$\left. + \left(1 + \frac{ip}{q}\right) \left(\sin(qa_1) - \frac{q}{p}\cos(qa_1)\right) e^{-pl} \right]. \tag{52.7}$$

According to the rules discussed in the preceding section all the physical characteristics of the s-states can be determined from $F_0(q)$. It was shown, for instance, that for determining the half-life of the disintegrating states as well as the velocity of the emitted α-particles the roots of the equation

$F_0(q) = 0$ were wanted. According to (52.7) the equation yielding the disintegrating states is

$$\frac{p \sin(qa_1) + q \cos(qa_1)}{p \sin(qa_1) - q \cos(qa_1)} = -\frac{q + ip}{q - ip} e^{-2pl}. \tag{52.8}$$

This transcendental equation can be solved only by numerical (or graphical) methods. The position of the solution in the complex q-plane for different values of V_0 is seen in Fig. 29. As can be read from the figure for increasing V_0 the root $q = -k_0 + i\kappa$ gets nearer and nearer to the origin, that is the values both of k_0 and of κ become smaller and smaller. And so from (51.22) and (51.25) it follows that an increase in V_0 reduces the kinetic energy E_0 of the α-particles and increases the half-life T. Thus to a longer

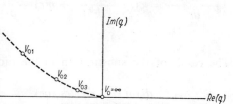

FIG. 29. Disintegrating states for the case of a rectangular potential barrier. $V_{01} < V_{02} < V_{03} < \ldots$ are the increasing values of the potential barrier

half-life a small kinetic energy corresponds and vice versa, in accordance with the Geiger–Nuttall law.

53. The Method of Integral Equations

As discussed in Section 51 in the case of a central potential field the basic problem is always the solution of the radial Schrödinger equation and the determination of the coefficient $F_l(q)$ determining the asymptotic behaviour of the physical solution. If $F_l(q)$ is known both the energy of the bound states as well as the scattering cross-sections can be determined by means of simple rules. The problem of solving the radial Schrödinger equation can be accelerated by means of the so-called *method of integral equations*, especially if electronic computers are available. In the following this method will be discussed.

We assume the physical solution $u_l(q, r)$ in the form:

$$u_l(q, r) = \xi(r) j_l(qr) + \eta(r) n_l(qr). \tag{53.1}$$

Here ξ and η are as yet unknown functions, j_l and n_l are the expressions given in (50.3). Between ξ and η the following relation is still imposed:

$$\frac{d\xi}{dr} j_l(qr) + \frac{d\eta}{dr} n_l(qr) = 0. \tag{53.2}$$

173

Let us substitute in the Schrödinger equation (51.3). Due to (53.2) the differentiating of (51.3) yields

$$\frac{du_l(q, r)}{dr} = \xi(r) \frac{dj_l(qr)}{dr} + \eta(r) \frac{dn_l(qr)}{dr}, \qquad (53.3)$$

and

$$\frac{d^2u_l(q, r)}{dr^2} = \xi(r) \frac{d^2j_l(qr)}{dr^2} + \eta(r) \frac{d^2n_l(qr)}{dr^2} + \frac{d\xi}{dr} \frac{dj_l(qr)}{dr} + \frac{d\eta}{dr} \frac{dn_l(qr)}{dr}.$$

$$(53.4)$$

The result of the substitution in the radial Schrödinger equation is

$$\frac{d\xi}{dr} \frac{dj_l(qr)}{dr} + \frac{d\eta}{dr} \frac{dn_l(qr)}{dr}$$

$$= - \xi(r) \left[\frac{d^2j_l(qr)}{dr^2} + \left(q^2 - U(r) - \frac{l(l+1)}{r^2} \right) j_l(qr) \right]$$

$$- \eta(r) \left[\frac{d^2n_l(qr)}{dr^2} + \left(q^2 - U(r) - \frac{l(l+1)}{r^2} \right) n_l(qr) \right]. \qquad (53.5)$$

If we remember that $j_l(qr)$ and $n_l(qr)$ satisfy the differential equation (50.1), i.e. the radial Schrödinger equation valid for the case without an exterior potential, then,

$$\frac{d^2j_l(qr)}{dr^2} + \left(q^2 - \frac{l(l+1)}{r^2} \right) j_l(qr) = 0,$$

$$(53.6)$$

$$\frac{d^2n_l(qr)}{dr^2} + \left(q^2 - \frac{l(l+1)}{r^2} \right) n_l(qr) = 0.$$

Due to this (53.5) becomes

$$\frac{d\xi}{dr} \frac{dj_l(qr)}{dr} + \frac{d\eta}{dr} \frac{dn_l(qr)}{dr} = U(r) u_l(q, r). \qquad (53.7)$$

Equations (53.2) and (53.7) can be regarded as a system of simultaneous equations from which the derivatives $d\xi/dr$ and $d\eta/dr$ can be derived. The relation

$$\frac{dn_l(qr)}{dr} j_l(qr) - \frac{dj_l(qr)}{dr} n_l(qr) = q \qquad (53.8)$$

174

valid for the Bessel functions (50.3) may be used for solving the system of simultaneous equations.

From this we obtain

$$\frac{d\xi}{dr} = -\frac{1}{q} n_l(qr)\, U(r)\, u_l(q,r)\,,$$

$$(53.9)$$

$$\frac{d\eta}{dr} = \frac{1}{q} j_l(qr)\, U(r)\, u_l(q,r)\,.$$

Integrating these equations we get ξ and η. We still must know the initial conditions of ξ and η. The behaviour of functions j_l and n_l for $r \to 0$ can be obtained from (50.6):

$$j_l(qr) \approx \frac{(qr)^{l+1}}{(2l+1)!!} \quad \text{and} \quad n_l(qr) \approx -\frac{(2l-1)!!}{(qr)^l}\,, \quad \text{if } r \to 0. \quad (53.10)$$

(53.1) satisfies the condition (51.4) for the physical solution if

$$\xi(0) = \frac{(2l+1)!!}{q^{l+1}} \quad \text{and} \quad \eta(0) = 0\,. \quad (53.11)$$

Taking these into consideration the integration of (53.9) yields

$$\xi(r) = \frac{1}{q}\left[\frac{(2l+1)!!}{q^l} - \int_0^r n_l(qr')\, U(r')\, u_l(q,r')\, dr'\right],$$

$$(53.12)$$

$$\eta(r) = \frac{1}{q}\int_0^r j_l(qr')\, U(r')\, u_l(q,r')\, dr'\,.$$

Substituting these expressions back in (53.1) the desired integral equation is reached:

$$u_l(q,r) = \frac{1}{q}\left[\frac{(2l+1)!!}{q^l} j_l(qr) + \int_0^r \{n_l(qr)j_l(qr') - j_l(qr)n_l(qr')\}\, U(r')\, u_l(q,r')\, dr'\right].$$

$$(53.13)$$

This is an inhomogeneous integral equation of the Volterra type; solving it $u_l(qr)$ can be determined unambiguously.

Now let us investigate the asymptotic behaviour of u_l for the limiting case of $r \to \infty$. The asymptotic form of the functions j_l and n_l are determined by the expressions (50.10)

$$j_l(qr) \approx \sin\left(qr - \frac{l\pi}{2}\right)$$

$$, \quad \text{if } r - \infty. \qquad (53.14)$$

$$n_l(qr) \approx - \cos\left(qr - \frac{l\pi}{2}\right)$$

In the range of very large r the upper limit of integration of (53.13) may be taken to be ∞, as $U(r)$ becomes negligibly small for large r values. Therefore the asymptotic form of (53.13) is

$$u_l(q, r) \approx \frac{1}{q}\left[\frac{(2l+1)!!}{q^l} - \int_0^\infty n_l(qr')\, U(r')\, u_l(q, r')\, dr'\right] \sin\left(qr - \frac{l\pi}{2}\right)$$

$$- \frac{1}{q}\int_0^\infty j_l(qr')\, U(r')\, u_l(q, r')\, dr' \cdot \cos\left(qr - \frac{l\pi}{2}\right), \quad \text{if } r \to \infty. \quad (53.15)$$

By using the Euler relations this may be put also in the following form:

$$u_l(q, r) \approx \frac{1}{2iq}\left[\frac{(2l+1)!!}{q^l} - \int_0^\infty (ij_l(qr') + n_l(qr'))\, U(r')\, u_l(q, r')dr'\right] e^{i(qr - l\pi/2)}$$

$$- \frac{1}{2iq}\left[\frac{(2l+1)!!}{q^l} + \int_0^\infty (ij_l(qr') - n_l(qr'))\, U(r')\, u_l(q, r')\, dr'\right] e^{-i(qr - l\pi/2)}.$$

$$(53.16)$$

This asymptotic behaviour corresponds to (51.9); the comparison yields the expression of $F_l(q)$ as well:

$$F_l(q) = e^{-il\pi/2}\left[\frac{(2l+1)!!}{q^l} - \int_0^\infty (ij_l(qr) + n_l(qr))\, U(r)\, u_l(q, r)\, dr\right].$$

$$(53.17)$$

Thus we reached our goal: an integral expression has been obtained for $F_l(q)$. Thus the course of applying the method of integral equations is as follows. First the Volterra type integral equation (53.13) is solved, then substituting the solution $u_l(q, r)$ in (53.17) $F_l(q)$ is computed. We mention that although principally the integration is performed from 0 till ∞, in practice it is sufficient to integrate for that region where $U(r)$ differs from zero. The method of integral equations finds application mostly in scattering calculus. Equation (53.17) implies, of course, the solution of the scattering problem as well, the value of the phase constant δ_l is given by rule (51.17):

$$\delta_l(k) = -\arctan \frac{\int_0^{\infty} j_l(kr)\, U(r)\, u_l(k, r)\, dr}{\dfrac{(2l + 1)!!}{k^l} - \int_0^{\infty} n_l(kr)\, U(r)\, u_l(k, r)\, dr}. \qquad (53.18)$$

54. Scattering by a Sphere with Rigid Walls

In the preceding sections the general theory of scattering has been dealt with. During the discussions it was supposed tacitly that, on the one hand, the potential V has no strong singularities, while, on the other hand, far from the centre of the field the potential vanishes rapidly. For most of the simple potentials these conditions hold. There are, however, two practical potentials showing an exception. One of them is the case of the hard sphere, the other the Coulomb potential. The latter will be discussed in the next section, now we deal with the scattering by a hard sphere.

Let us denote the radius of the sphere by a. The potential is

$$V(r) = +\infty, \quad \text{if } r < a,$$
$$V(r) = 0, \quad \text{if } r > a. \qquad (54.1)$$

The scattered particle cannot penetrate into the sphere, thus the wave function ψ must vanish on the surface of the sphere, and is zero inside it. The wave function satisfies this condition only, if it is valid for each partial wave so that

$$u_l(r) = 0, \quad \text{if } r \leq a. \qquad (54.2)$$

This contradicts the initial condition (49.5) of the physical solution. This is obvious, as when deducing (49.5) it was supposed that for $r \rightarrow 0$ the potential energy can be neglected in the Schrödinger equation besides the term $l(l + 1)/r^2$. In the case of the potential (54.1) the situation is, however, quite different. Here the boundary conditions (54.2) take the role of (49.5).

In the region outside the sphere, as the potential here is zero, the form of the radial Schrödinger equation is

$$\frac{d^2u_l}{dr^2} + \left(k^2 - \frac{l(l+1)}{r^2}\right)u_l = 0, \quad \text{if } r \geq a. \tag{54.3}$$

This is the same as the second equation (50.1); its general solution has already been put under (50.4):

$$u_l(r) = bj_l(kr) + cn_l(kr), \quad \text{if } r \geq a. \tag{54.4}$$

According to what was said before this has to vanish on the surface of the sphere, i.e. at $r = a$. From this the ratio of the constants of integration b and c is:

$$\frac{c}{b} = -\frac{j_l(ka)}{n_l(ka)}. \tag{54.5}$$

The investigation of the behaviour of the solution (54.5) for $r \to \infty$ yields the phase constant δ_l. By means of the asymptotic formulae of (53.14) we find that

$$u_l(r) \approx b \sin\left(kr - \frac{l\pi}{2}\right) - c \cos\left(kr - \frac{l\pi}{2}\right), \quad \text{if } r \to \infty. \tag{54.6}$$

Or, in a slightly different form,

$$u_l(r) \approx (b^2 + c^2)^{\frac{1}{2}} \sin\left(kr - \frac{l\pi}{2} - \arctan\frac{c}{b}\right), \quad \text{if } r \to \infty. \tag{54.7}$$

Comparing this with (49.7) it can be seen that

$$\delta_l = -\arctan\frac{c}{b},$$

so that

$$\delta_l = \arctan\frac{j_l(ka)}{n_l(ka)}. \tag{54.8}$$

The scattering problem is essentially solved by determining the phase constant. Figure 30 shows the differential scattering cross-section for $ka = 20$, i.e. if the bombarding energy is $E = 50h^2/\pi^2 m_0 a^2$.

The total scattering cross-section is obtained by substituting the phase constants of (54.7) in (49.16)

$$\sigma = \frac{4\pi}{k^2} \sum_{l=0}^{\infty} \frac{(2l+1)j_l^2(ka)}{j_l^2(ka) + n_l^2(ka)}. \tag{54.9}$$

At very low bombarding energies, i.e. if $ka \ll 1$, the scattering problem becomes very simple. In this case for very small arguments the approximations of (53.10) can be used. This yields for the phase constant of the scattering

$$\delta_l \approx - \arctan \frac{(ka)^{2l+1}}{(2l-1)!! \, (2l+1)!!} . \tag{54.10}$$

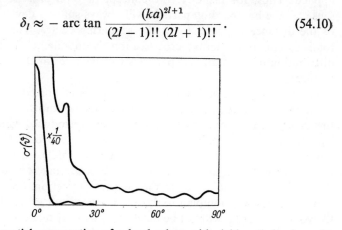

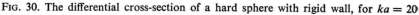

FIG. 30. The differential cross-section of a hard sphere with rigid wall, for $ka = 20$

With increasing l the phase constant of scattering decreases rapidly. The dominating phase constant is the one belonging to the s-wave:

$$\delta_0 \approx - \arctan (ka) \approx -ka . \tag{54.11}$$

The total cross-section of scattering in this case is

$$\sigma = 4\pi a^2 . \tag{54.12}$$

This is equal to the fourfold of the geometric cross-section.

55. Coulomb Scattering

In the case of Coulomb scattering the interaction of the scattering centre and the bombarding particle is described by the Coulomb potential:

$$V = \frac{Z_1 Z_2 e^2}{r} . \tag{55.1}$$

Here $Z_1 e$ is the electric charge of the scattering centre, $Z_2 e$ is that of the bombarding particle. With increasing r the Coulomb potential vanishes only very slowly, therefore its range can be taken as infinite. This is the reason why the Coulomb scattering is to be regarded as an exception for which the scattering theory developed in the preceding sections cannot be applied. Thus, for instance, the concept of total cross-section is void of sense, as the bombarding particle will always suffer a deviation irrespective of the distance it passes by at the scattering centre. This means that the total Coulomb scattering cross-section is infinite. Only the differential distribution of the scattered particles can be used for the description of the scattering.

The investigation of Coulomb scattering is of great importance both in atomic and nuclear physics. Several methods are known for the solution of this problem. Here only a special one will be introduced which is not linked with the method of partial waves and can be used only if for every r the potential is described by the Coulomb law of (55.1). In the physics of the nuclei the so-called anomalous Coulomb scattering is known where near the origin the potential, due to the nuclear interaction, deviates from (55.1) and becomes Coulombian only at large distances r. The anomalous Coulomb scattering will not be dealt with, we only mention that the method of partial waves can be generalized in such a form that it can be applied also to the case of long-range Coulomb forces.

The Schrödinger equation to be solved is

$$\nabla^2 \psi + \left(k^2 - \frac{2nk}{r} \right) \psi = 0 , \tag{55.2}$$

where k is the wave number defined by (48.3), and

$$n = \frac{4\pi^2 Z_1 Z_2 e^2 m_0}{h^2 k} . \tag{55.3}$$

n is a dimensionless quantity, proportional to the product $Z_1 Z_2$ characterizing the strength of the Coulomb interaction, and it is inversely proportional to k, that is to the initial velocity of the bombarding particle $v_0 = hk/2\pi m_0$.

The solution of the Schrödinger equation is assumed to be

$$\psi = e^{ikz} g(r - z) . \tag{55.4}$$

This yields for g the following second-order linear differential equation:

$$t \frac{d^2 g}{dt^2} + (1 - ikt) \frac{dg}{dt} - nk g(t) = 0 , \tag{55.5}$$

where $t = r - z$. The solution of g is required in the form of a power series

$$g(t) = \sum_{p=0}^{\infty} (a_p t^p / p!) .$$

After substitution and arranging the expression according to the powers of t we obtain the equation

$$\sum_{p=0}^{\infty} [(p + 1) a_{p+1} - k(n + ip) a_p] t^p / p! = 0 .$$

This can be valid for every value of t only, if the coefficient of each power of t is equal to zero. This yields for the coefficients a_p the following recurrence relation:

$$a_{p+1} = k \frac{n + ip}{p + 1} a_p . \tag{55.6}$$

From this

$$a_p = \frac{n(n + i)(n + 2i) \ldots [n + (p - 1)i]}{p!} k^p a_0 . \tag{55.7}$$

Thus the solution of the differential equation (55.5) is

$$g(t) = a_0 \left[1 + \frac{n}{1!} \frac{kt}{1!} + \frac{n(n + i)}{2!} \frac{(kt)^2}{2!} + \frac{n(n + i)(n + 2i)}{3!} \frac{(kt)^3}{3!} + \ldots \right] , \tag{55.8}$$

where a_0 is a normalizing constant which is as yet undetermined. The power series in square brackets is a special form of the confluent hypergeometric series known from mathematics:

$$F(a, b; x) = 1 + \frac{a}{b} \frac{x}{1!} + \frac{a(a + 1)}{b(b + 1)} \frac{x^2}{2!} + \frac{a(a + 1)(a + 2)}{b(b + 1)(b + 2)} \frac{x^3}{3!} + \ldots , \tag{55.9}$$

namely in (55.8) the series $F(-in, 1; ikt)$ is present.

The Schrödinger equation thus actually has a regular solution of the form (55.4); the function is

$$\psi = a_0 e^{ikz} F(-in, 1; ik(r - z)) . \tag{55.10}$$

Let us investigate the behaviour of the wave function ψ at large distances from the scattering centre. According to the theory of the confluent hypergeometric series the sum of (55.9) can be described for $|x| \to \infty$ by the asymptotic formula

$$F(a, b; x) \approx \frac{\Gamma(b)}{\Gamma(b - a)}(-x)^{-a} + \frac{\Gamma(b)}{\Gamma(a)}e^x x^{a-b}.$$

In the special case we are interested in

$$F(-in, 1; ikt) \approx \frac{e^{n\pi/2}}{\Gamma(1 + in)}\{(kt)^{in} - ne^{i(kt+2\eta_0)}(kt)^{-(1+in)}\}, \quad (55.11)$$

where

$$e^{2i\eta_0} = \frac{\Gamma(1 + in)}{\Gamma(1 - in)}.$$

By using this the asymptotic form for the function ψ in the case of $|r - z| \to \infty$, i.e. (apart from the z-axis) far from the origin,

$$\psi \approx \frac{a_0 e^{n\pi/2}}{\Gamma(1 + in)}\left\{e^{i[kz+n \ln k(r-z)]} - \frac{n}{k(r - z)}e^{i[kr-n \ln k(r-z)+2\eta_0]}\right\}$$

$$= \frac{a_0 e^{n\pi/2}}{\Gamma(1 + in)}\left\{e^{i(kz+n \ln k(r-z))} + f_{\text{Coul}}(\vartheta)\frac{e^{i[kr-n \ln kr]}}{r}\right\}, \quad (55.12)$$

where

$$f_{\text{Coul}}(\vartheta) = -\frac{ne^{i[-n \ln (1- \cos \vartheta)+\eta_0]}}{k(1 - \cos \vartheta)}. \quad (55.13)$$

The asymptotic behaviour of this wave function satisfies the requirements set in Section 48: it consists of an incident and a scattered wave. Comparing (55.12) and (48.5) a very interesting phenomenon can be observed. The long-range Coulomb force distorts both the incident and the scattered wave with a phase constant changing logarithmically. Also classical mechanics explains the case of this distortion. Let us investigate all the hyperbolic orbits, one asymptote of which goes in the direction of the incident particles, i.e. which are parallel with the negative z-axis. The surfaces perpendicular to the orbits corresponding to the wave fronts of the matter waves are not even for $z \to -\infty$, $z = $ const. planes but are the planes with

$$z + \frac{Z_1 Z_2 e^2}{m_0 v_0^2}\ln k(r - z) = \text{const}.$$

Accordingly the wave function describing the incident particles is

$$e^{ik[z + (Z_1 Z_2 e^2/m_0 v_0^2) \ln k(r-z)]}.$$

Taking (55.3) into consideration this is just the first term of (55.12). The distortion of the scattered spherical wave can be explained in a similar way. The amplitude of the scattered spherical wave is given by (55.13); from this the differential cross-section of the Coulomb scattering can be calculated according to rule (48.6):

$$\sigma(\vartheta) = |f_{\text{Coul}}(\vartheta)|^2 = \frac{n^2}{k^2(1 - \cos \vartheta)^2} \, .$$

Substituting here the expressions of k and n, the following end result is reached:

$$\sigma(\vartheta) = \frac{Z_1^2 Z_2^2 e^4}{4m_0^2 v_0^4 \sin^4 \left(\dfrac{\vartheta}{2}\right)} \, . \tag{55.14}$$

This is the well-known Rutherford formula. Its noteworthy property is that it does not contain the Planck constant h, the characteristic constant of quantum physics. This is connected with the fact that the Rutherford formula can be deduced in classical mechanics as well. Even Rutherford deduced the formula by the help of classical mechanics during the investigation of the scattering of α-particles by nuclei. Expression (2.1) describing the scattering of α-particles is a direct consequence of the Rutherford formula.

Finally let us investigate the probability that the particle scattered in a Coulomb field reaches the centre of the field. From the interpretation of the wave function it follows that this probability is given by $|\psi(0)|^2$. According to (55.10) the value of the wave function at the origin equals with the normalizing constant a_0

$$\psi(0) = a_0 \, ,$$

as the value of the hypergeometric series F at the origin is equal to unity. The value of the constant a_0 can be determined from the asymptotic form of the wave function. Comparing eqs. (55.12) and (48.5) we can see that

$$\frac{a_0 e^{n\pi/2}}{\Gamma(1 + in)} = N = \sqrt{\frac{2\pi m_0 s_0}{hk}} \, .$$

By using the relation

$$k = \frac{2\pi m_0 v_0}{h}$$

we find

$$a_0 = \sqrt{\frac{s_0}{v_0}}\, e^{-n\pi/2}\, \Gamma(1 + in)\,.$$ (55.15)

Finally the required probability density is

$$|\,\psi(0)\,|^2 = \frac{s_0}{v_0}\, \frac{2\pi n}{e^{2\pi n} - 1}\,.$$ (55.16)

If the charge of the scattering centre is zero, i.e. if $n = 0$, then, of course, no scattering takes place. In this case the bombarding particles fill the space with a uniform density $\rho_0 = s_0/v_0$. The factor

$$G(n) = \frac{2\pi n}{e^{2\pi n} - 1}$$

occurring in (55.16) is a multiplier accounting for the increase (or decrease) of the density of incident particles at the centre of the potential compared to the average density ρ_0. If $|\,n\,| \ll 1$, i.e. if the Coulomb interaction is weak or the velocity of the bombarding particles is very high, then approximately

$$G(n) \approx 1 - \pi n\,.$$

That means that for a repulsive interaction ($n > 0$) the G factor is somewhat smaller and for an attractive interaction ($n < 0$) somewhat higher than unity. In the limiting case of $|\,n\,| \gg 1$, i.e. for small bombarding velocities and strong interactions, G can be put in the form of the following approximate expressions:

$$G(n) \approx 2\pi n e^{-2\pi n}\,, \quad \text{if } n > 0\,,$$
$$G(n) \approx 2\pi\,|\,n\,|\,, \quad \text{if } n < 0\,.$$

The second formula shows that in the case of an attractive interaction G increases proportional to $|\,n\,|$. The first expression, valid for repulsive interactions, contains the so-called Gamow factor:

$$e^{-2\pi n} = e^{-4\pi^2 Z_1 Z_2 e^2/h v_0}\,.$$ (55.17)

The reaction between two positively charged nuclei takes place (supposing that the radii of the nuclei can be regarded small enough) only then, if the distance between the nuclei decreases to zero. The probability of this is, as seen, proportional to $G(n)$. For small bombarding energies the order of magnitude of the probability of nuclear reactions is determined primarily by the Gamow factor (55.17).

CHAPTER 6

The Simplest Approximate Methods and Their Applications

56. Perturbation Calculus

In several cases the Schrödinger equation of the atomic systems cannot be solved exactly due to the occurring mathematical difficulties. Namely, in the case of more complicated systems the Schrödinger equation becomes such a partial differential equation that the special functions investigated in mathematical analysis are not sufficient for solving it. In such cases the methods enabling the approximate solutions of the Schrödinger equation become important. In this section the simplest method, the perturbation calculus will be introduced. This method can be applied if the Schrödinger equation of the actual system differs only slightly from that of a simple system with solutions already known.

The energy operator of the investigated system is accordingly

$$H = H_0 + \mathcal{H}, \tag{56.1}$$

where H_0 is a simple energy operator, the corresponding Schrödinger equation of which is

$$H_0 \psi_k^0 = E_k^0 \psi_k^0. \tag{56.2}$$

The eigenvalues E_k^0 and eigenfunctions ψ_k^0 of this equation are taken as known. The second term in H, $\mathcal{H}$ is taken as a small perturbation. The Schrödinger equation to be solved is

$$H\psi \equiv (H_0 + \mathcal{H}) \psi = E\psi. \tag{56.3}$$

As the effect of the perturbation $\mathcal{H}$ is supposed to be small, it is assumed that the eigenvalue E will lie near one of the eigenvalues E_k^0 of the operator H_0. Therefore let us assume for E the following form:

$$E = E_k^0 + w, \tag{56.4}$$

where the correction w denotes the energy change due to the perturbation. ψ is also supposed to lie near the eigenfunction corresponding to E_k^0. Here, however, a distinction has to be made whether E_k^0 is a non-degenerate eigenvalue, i.e. only a single eigenfunction corresponds to it, or whether it is a degenerate one with more than one corresponding eigenfunctions. We begin with an examination of the non-degenerate eigenvalues. In this case ψ will differ from ψ_k^0 only by a small correction

$$\psi = \psi_k^0 + \phi . \tag{56.5}$$

Let us substitute in the Schrödinger equation (56.3) the expressions (56.4) and (56.5). Taking into account that ψ_k^0 satisfies the Schrödinger equation (56.2) corresponding to H_0, we find that

$$H_0\phi + \mathcal{H}\psi_k^0 + \mathcal{H}\phi = E_k^0\phi + w\psi_k^0 + w\phi . \tag{56.6}$$

By expanding the correction ϕ into a power series in powers of the unperturbed eigenfunctions ψ_n^0

$$\phi = \sum_n a_n\psi_n^0 , \tag{56.7}$$

and substituting this in (56.6) we get

$$\sum_n a_n E_n^0\psi_n^0 + \mathcal{H}\psi_k^0 + \sum_n a_n\mathcal{H}\psi_n^0 = w\psi_k^0 + (E_k^0 + w)\sum_n a_n\psi_n^0 . \tag{56.8}$$

On the left-hand side in the first term of the expression the eigenvalue equation (56.2) could be used again. We multiply (56.8) first by ψ_k^{0*} then integrate over all configuration space. With regard to the normalizing and orthogonality relations (26.7) and (26.8) the result with the introduction of the simplifying notation

$$\mathcal{H}_{kn} = \int \psi_k^{0*}\mathcal{H}\psi_n^0 \, dv \tag{56.9}$$

is

$$\mathcal{H}_{kk} + \sum_n a_n\mathcal{H}_{kn} = (1 + a_k)\, w . \tag{56.10}$$

This equation enables the expansion of the energy correction w by means of the coefficients a_n of the series expansion. Introducing the notation

$$b_n = \frac{a_n}{1 + a_k} . \tag{56.11}$$

from (56.10) we obtain the following expression for the energy correction term

$$w = \mathcal{H}_{kk} + \sum_n{}' \mathcal{H}_{kn} b_n \tag{56.12}$$

where the comma on the Σ denotes that the k-term of the summation is to be omitted. By means of (56.12) the determination of the energy correction

186

has been reduced to the determination of the coefficients a_n (or equivalently of the eigenfunction correction). For calculating the coefficients a_n we multiply (56.8) throughout by ψ_m^{0*} ($m \neq k$) and integrate. The result is

$$a_m E_m^0 + \mathcal{H}_{mk} + \sum_n a_n \mathcal{H}_{mn} = (E_k^0 + w) a_m .$$

By a simple calculation from this we may deduce for the constants b_m the equation

$$b_m = \frac{\mathcal{H}_{mk}}{E_k^0 - E_m^0 + w} + \sum_n{}' \frac{\mathcal{H}_{mn}}{E_k^0 - E_m^0 + w} b_n . \qquad (56.13)$$

The solution of this equation by the method of successive approximations is simple. As a starting point b_n is regarded to be equal to zero on the right-hand side, then

$$b_m = \frac{\mathcal{H}_{mk}}{E_k^0 - E_m^0 + w} .$$

As the next step this expression for the first approximation of b_n is used in the right-hand side; then we receive

$$b_m = \frac{\mathcal{H}_{mk}}{E_k^0 - E_m^0 + w} + \sum_n{}' \frac{\mathcal{H}_{mn}\mathcal{H}_{nk}}{(E_k^0 - E_m^0 + w)(E_k^0 - E_n^0 + w)} .$$

Continuing this method for b_m an expression consisting of an infinite series is obtained

$$b_m = \frac{\mathcal{H}_{mk}}{E_k^0 - E_m^0 + w} + \sum_n{}' \frac{\mathcal{H}_{mn}\mathcal{H}_{nk}}{(E_k^0 - E_m^0 + w)(E_k^0 - E_n^0 + w)}$$

$$+ \sum_{n,l}{}' \frac{\mathcal{H}_{mn}\mathcal{H}_{nl}\mathcal{H}_{lk}}{(E_k^0 - E_m^0 + w)(E_k^0 - E_n^0 + w)(E_k^0 - E_l^0 + w)} + \dots . \qquad (56.14)$$

If this is finally substituted in (56.12) a transcendental equation is found for the energy correction

$$w = S(w) \equiv \mathcal{H}_{kk} + \sum_m{}' \frac{\mathcal{H}_{km}\mathcal{H}_{mk}}{E_k^0 - E_m^0 + w}$$

$$+ \sum_{m,n}{}' \frac{\mathcal{H}_{km}\mathcal{H}_{mn}\mathcal{H}_{nk}}{(E_k^0 - E_m^0 + w)(E_k^0 - E_n^0 + w)} + \dots . \qquad (56.15)$$

It has to be emphasized that up to now no approximating step has been used; by solving the above transcendent equation the exact eigenvalues of (56.3) would be found. In the case of small perturbations $\mathcal{H}$, however, the infinite series standing on the right-hand side of (56.15) can be approximated by its first few terms; the Nth term contains the product of N matrix elements $\mathcal{H}$ and thus becomes for larger N extremely small. In the first approximation of perturbation calculus only the first term is taken into consideration, then

$$w = \mathcal{H}_{kk} = \int \psi_k^{0*} \, \mathcal{H} \psi_k^0 \, dv \, . \tag{56.16}$$

The energy correction is equal to the quantum-mechanical mean value of the perturbation $\mathcal{H}$ taken in the ψ_k^0 state.

Thus the first-order correction to the energy is the mean value of the perturbation averaged over the unperturbed wave function. In the second approximation two terms are already taken into consideration from (56.15); but in the denominator of the second correction term w may be neglected with respect to the energy difference $E_k^0 - E_m^0$:

$$w = \mathcal{H}_{kk} + \sum_m{}' \, \frac{\mathcal{H}_{km} \mathcal{H}_{mk}}{E_k^0 - E_m^0} \, . \tag{56.17}$$

Let us now deal with the correction ϕ of the eigenfunction. In first approximation only the first term is taken into consideration in (56.14) and also here $w = 0$ is put in the denominator

$$b_m = \frac{\mathcal{H}_{mk}}{E_k^0 - E_m^0} \, . \tag{56.18}$$

According to (56.11) $a_m = (1 + a_k) \, b_m$, and the correction of the eigenfunction is

$$\phi = \sum_m a_m \psi_m^0 = a_k \psi_k^0 + (1 + a_k) \sum_m{}' \, b_m \psi_m^0 \, .$$

If the approximate expression (56.18) is written into this equation, then the following formula is gained for the total $\psi = \psi_k^0 + \phi$ wave function

$$\psi = (1 + a_k) \left[\psi_k^0 + \sum_m{}' \, \frac{\mathcal{H}_{mk}}{E_k^0 - E_m^0} \, \psi_m^0 \right] . \tag{56.19}$$

The value of the coefficient $(1 + a_k)$ can be determined from the normalizing condition of (26.7). If we do not strive for determining a normalized eigenfunction ψ, then we can simply put $a_k = 0$ and the eigenfunction will be the expression standing in square brackets in (56.19).

188

Let us now deal with the case of degenerate eigenvalues. If the number of linearly independent wave functions corresponding to the eigenvalue E_k^0 is f,

$$\psi_{k1}^0, \psi_{k2}^0, \ldots, \psi_{kf}^0 \tag{56.20}$$

then it is not possible to tell without any further investigation which linear combination of the functions $\psi_{k\alpha}^0$ will be the zeroth-order approximation of the function ψ corresponding to $H = H_0 + \mathcal{H}$. It is obvious that instead of (56.5) the expression

$$\psi = \sum_{\alpha=1}^{f} c_\alpha \psi_{k\alpha}^0 + \phi \tag{56.21}$$

has to be written, where the c_α-s are still unknown coefficients. Substituting in the Schrödinger equation $(H_0 + \mathcal{H})\psi = (E_k^0 + w)\psi$ instead of (56.6) one obtains the following equation:

$$H_0\phi + \sum_{\alpha=1}^{f} c_\alpha \mathcal{H}\psi_{k\alpha}^0 + \mathcal{H}\phi = E_k^0\phi + w \sum_{\alpha=1}^{f} c_\alpha \psi_{k\alpha}^0 + w\phi. \tag{56.22}$$

For the sake of simplicity the energy correction w will be determined only in the first approximation of perturbation calculus. In this case from (56.22) the expressions $\mathcal{H}\phi$ and $w\phi$, as second-order corrections, can be neglected. We multiply the remaining equation by $\psi_{k\beta}^{0*}$ and integrate over all space. If similarly to (56.9) the notation

$$\mathcal{H}_{\beta\alpha}^{(k)} = \int \psi_{k\beta}^{0*} \mathcal{H}\psi_{k\alpha}^0 \, dv \tag{56.23}$$

is introduced, and, on the other hand, we consider that the relations (26.7) and (26.8)

$$\int \psi_{k\alpha}^{0*} \psi_{k\alpha}^0 \, dv = 1 \quad \text{and} \quad \int \psi_{k\beta}^{0*} \psi_{k\alpha}^0 \, dv = 0 , \quad \text{if } \beta \neq \alpha$$

are valid for (56.20), then the following result is found:

$$\sum_{\alpha=1}^{f} (\mathcal{H}_{\beta\alpha}^{(k)} - w\delta_{\beta\alpha}) c_\alpha = 0 . \tag{56.24}$$

Here $\delta_{\beta\alpha}$ is the so-called Kronecker δ symbol, its value is 1, if $\beta = \alpha$, and 0, if $\beta \neq \alpha$. Equation (56.24) is a system of simultaneous linear homogeneous equations. The coefficients c_α can be determined from it. The condition of solubility is that the determinant of the system of equations should vanish:

$$\begin{vmatrix} \mathcal{H}_{11}^{(k)} - w & \mathcal{H}_{12}^{(k)} & \dots & \mathcal{H}_{1f}^{(k)} \\ \mathcal{H}_{21}^{(k)} & \mathcal{H}_{22}^{(k)} - w & \dots & \mathcal{H}_{2f}^{(k)} \\ \cdot & \cdot & & \cdot \\ \cdot & \cdot & & \cdot \\ \cdot & \cdot & & \cdot \\ \mathcal{H}_{f1}^{(k)} & \mathcal{H}_{f2}^{(k)} & \dots & \mathcal{H}_{ff}^{(k)} - w \end{vmatrix} = 0. \quad (56.25)$$

The value of the determinant gives an algebraic equation for w of the order f, which is called a secular equation. Its solutions, the roots w_1, $w_2, \dots, w_f$ yield the corrections of the energy value E_k^0. If all the f roots are different, the degeneracy is lifted. As, according to our suppositions, the matrix elements $\mathcal{H}_{\beta\alpha}^{(k)}$ are small, the roots of the secular equation will be small as well. Thus due to the perturbation the degenerated energy value E_k^0 decomposes into f levels

$$E_k^0 + w_1, \qquad E_k^0 + w_2, \dots, \qquad E_k^0 + w_f \qquad (56.26)$$

lying near each other. If not all the roots w are different, the energy splitting is only partial.

To each energy correction w a system of solutions of (56.24) belongs. Through the computation of the coefficients c_α it becomes possible to obtain also the first term of (56.21):

$$\psi^0 = \sum_{\alpha=1}^{f} c_\alpha \psi_{k\alpha}^0 ; \qquad (56.27)$$

this is the correct starting function.

In the following sections simple examples of the perturbation calculus will be demonstrated.

57. The Anharmonic Linear Oscillator

The potential energy of a mass point vibrating along a straight line can be put only in a very idealized case as $V = \dfrac{1}{2} kx^2$, the potential of the harmonic oscillator. A more realistic description of the vibration is reached, if the expression of ψ is augmented by an anharmonic, fourth power term in x:

$$V = \frac{1}{2} kx^2 + \lambda x^4 ; \qquad (57.1)$$

If λ is small, then the extra term $\mathcal{H} = \lambda x^4$ can be regarded besides the energy of the harmonic oscillator as a perturbation. The eigenvalues and eigenfunctions of the harmonic oscillator have been discussed in Section 32. The results are given in eqs. (32.22) and (32.25). Let us calculate the corrections of the energy levels $E_n^0 = h\nu\,(n + \frac{1}{2})$ by first-order perturbation calculus. According to (56.16)

$$w = \int_{-\infty}^{\infty} \psi_n^0 \mathcal{H} \psi_n^0\, dx = \frac{A_n^2}{\alpha^{3/2}} \int_{-\infty}^{\infty} e^{-\xi^2} H_n^2(\xi)\, \xi^4 d\xi , \qquad (57.2)$$

where the notation of Section 32 has been used.

This integral can be calculated by means of the Hermite polynomials of formula (32.24), the result is $3\sqrt{\pi}\,2^{n-2}\,n!\,(2n^2 + 2n + 1)$. Substituting expression (32.26) of the normalizing coefficient A_n in (57.2) we obtain

$$w = \frac{3h^2\lambda}{64\pi^4 m_0^2 \nu^2}\,(2n^2 + 2n + 1) . \qquad (57.3)$$

Our perturbation method is acceptable only if w is really only a small correction compared to the energy level E_n^0, i.e. if $w \ll h\nu\,(n + \frac{1}{2})$. From this we find for λ the condition:

$$\lambda \ll \frac{32\pi^4 m_0^2 \nu^3}{3h}\;\frac{2n + 1}{2n^2 + 2n + 1} . \qquad (57.4)$$

This means for the $n = 0$ ground level the condition: $\lambda \ll 32\pi^4 m_0^2 \nu^3/3h$. For the excited levels this limit has still to be multiplied by $2n + 1/2n^2 + 2n + 1$, a constant smaller than unity. For high n values this expression becomes very small. Thus the energy correction (57.3) can be applied only for the ground state and the low-lying excited states.

58. The Plane Rotator in an Electric Field

Let us place the plane rotator of a mass point m_0 and a radius a in a homogeneous electric field parallel to the plane of the rotator. If the electric charge of the rotator is denoted by e and the electric field strength by $\mathfrak{E}$, then the potential energy obviously is

$$\mathcal{H} = -e\mathfrak{E}\, a \cos \phi \, . \tag{58.1}$$

If the field strength is not too high, then this can be regarded as a perturbation besides the rotational energy of the rotator. The Schrödinger equation of the unperturbed rotator has been solved in Section 31:

$$\psi_n^0 = \frac{1}{\sqrt{2\pi}}\, e^{in\phi}\, , \quad E_n^0 = \frac{h^2 n^2}{8\pi^2 \Theta}\, , \tag{58.2}$$

where $\Theta = m_0 a^2$ is the moment of inertia of the rotator. Let us now produce the matrix elements of the perturbation energy $\mathcal{H}$ according to rule (56.9):

$$\mathcal{H}_{nm} = -\frac{e\mathfrak{E}a}{2\pi} \int_0^{2\pi} e^{-in\phi} \cos \phi \, e^{im\phi}\, d\phi = -\frac{1}{2}\, e\mathfrak{E}a(\delta_{m,n-1} + \delta_{m,n+1})\, ,$$

$$\tag{58.3}$$

where δ is again the Kronecker symbol. Knowing the matrix elements $\mathcal{H}_{nm}$ the energy correction w caused by the perturbation can be written down. In first approximation $w = \mathcal{H}_{nm}$, this is, however, according to (58.3) equal to zero; in first approximation also the energy of the perturbed rotator equals E_n^0. In second approximation the energy correction is given by (56.17), substituting in this the matrix element (58.3), we have that

$$w = \frac{e^2\mathfrak{E}^2 a^2}{4} \left(\frac{1}{E_n^0 - E_{n-1}^0} + \frac{1}{E_n^0 - E_{n+1}^0} \right) .$$

By the help of expression (57.2) of the unperturbed energy levels the end result is

$$w = \frac{4\pi^2 \Theta\, e^2\mathfrak{E}^2 a^2}{h^2} \cdot \frac{1}{4n^2 - 1} \, . \tag{58.4}$$

In the case of the ground state, i.e. for $n = 0$ this correction is negative, the external field reduces the energy of the rotator. For the excited states $(n \geq 1)$ w is positive, the perturbation lifts the energy levels.

192

59. The Effect of the Finite Size of the Nucleus on the Energy Levels of the Atom

In atomic physics the nucleus is generally regarded as a mass point with a charge Ze. This is, however, an idealization since it is known that the nuclei are of finite size. Let us regard the nucleus as a sphere of homogeneous charge distribution with a radius of R. The potential energy of the electron being at a distance r from the centre of the nucleus is

$$V(r) = -\frac{Ze^2}{R}\left(\frac{3}{2} - \frac{1}{2}\frac{r^2}{R^2}\right), \quad \text{if } r \le R,$$

$$V(r) = -\frac{Ze^2}{r}, \quad \text{if } r \ge R.$$

(59.1)

Subtracting the potential energy $V^0 = -Ze^2/r$ corresponding to the point-like nucleus, the perturbation due to the finite size of the nucleus is found:

$$\mathcal{H} = \frac{Ze^2}{R}\left(\frac{R}{r} - \frac{3}{2} + \frac{1}{2}\frac{r^2}{R^2}\right),$$

(59.2)

for $r \le R$, whereas its value is zero for $r \ge R$. We calculate the energy correction for the innermost, i.e. the $1s$ electron orbit. The corresponding wave function is obtained from (44.23) and (44.24) with the substitutions $n = 1$, $l = m = 0$:

$$\psi^0 = \left(\frac{Z^3}{\pi a_H^3}\right)^{\frac{1}{2}} e^{-Zr/a_H}.$$

In first-order approximation the value of the energy correction is

$$w = \int \psi^0 H \psi^0 \, dv = \frac{Z^4 e^2}{\pi a_H^3 R} \int_0^R \left(\frac{R}{r} - \frac{3}{2} + \frac{1}{2}\frac{r^2}{R^2}\right) e^{-2Zr/a_H} \, 4\pi r^2 \, dr.$$

By means of elementary calculations we find for the integral:

$$w = \frac{Z^2 e^2}{a_H}\left[1 - 3\frac{(4 + 4x + x^2)e^{-x} - (4 - x^2)}{x^3}\right],$$

(59.3)

where

$$x = \frac{2ZR}{a_H} = \frac{8\pi^2 Z m_0 e^2 R}{h^2}.$$

(59.4)

The energy of an electron moving in a $1s$ orbit around a point-like nucleus is

$$E_{1s}^0 = -Z^2 e^2/2a_H \, .$$

And so the relative energy change is

$$\eta = \frac{w}{|E_{1s}^0|} = 2\left[1 - 3\,\frac{(4 + 4x + x^2)e^{-x} - (4 - x^2)}{x^3}\right]. \qquad (59.5)$$

Let us consider, for example, the case of the lead atom. The atomic number of lead is $Z = 82$, the radius of its nucleus is $R = 7.1 \times 10^{-13}$ cm, from this the value of $x = 0.022$ is received. (For elements lighter than lead we gain x values even smaller by order of magnitudes.) For such small x values (59.5) can be approximated by the first non-vanishing term of its power series:

$$\eta = \frac{1}{5}\,x^2 \, . \qquad (59.6)$$

For the lead atom, taken as an example, the relative energy change is $\eta = 0.97 \times 10^{-4}$, thus the difference is only one-ten-thousandth of the energy calculated by supposing the nucleus being point-like. Spectroscopic investigations can, however, reveal even this energy correction. For the isotopes of a given element, where the atomic number Z is the same but the atomic weight, and thus the nuclear radii, are different, the energy corrections will be different and due to this the spectral lines will not coincide completely. This phenomenon is called isotope shift.

The situation is by orders of magnitudes different for the so-called muonic atoms. If a negative muon originating from cosmic rays or from an accelerator loses its kinetic energy (through the Compton effect), then during some time of its lifetime it can be captured in an orbit around a nucleus. The quantum theory of such a muonic atom is quite similar to that of a normal atom, the only, but important, difference is that the mass of the muon is 207 times bigger than that of the electron. In the example of a muon captured by a lead atom we find from (59.4) that $x = 207 \times 0.022 = 4.6$, and the corresponding relative energy correction is according to (59.5) $\eta = 0.90$. That means that the energy correction is 90 per cent. This means that the effects originating from the finite size of the nucleus are not to be regarded as perturbations: in the theory of the muon-atom the Schrödinger equation of the muon has to be solved in an exact manner with the potential (59.1). It becomes immediately obvious that the finite size of the nucleus has such a great effect on the muon, if we take into consideration, that due to the

207 times higher mass of the muon the Bohr radius a_H is 207 times smaller than that of an electron. The radius of the muon orbits lies in the order of the nuclear radius.

60. Spin–Orbit Interaction

The electrons moving in the atom get into a magnetic interaction with the electric field present in the atom, due to the magnetic moment of the electrons coupled to their spin. This is called *spin–orbit interaction*. It is known from electrodynamics that the magnetic energy of an electron, with its own magnetic moment μ_s, moving in an electrostatic field $\mathfrak{E}$, is

$$\frac{1}{c}\,\mu_s(\mathbf{v} \times \mathfrak{E})\,, \tag{60.1}$$

where $\mathbf{v}$ is the velocity of the electron. The electrostatic field of the atom is described by the function

$$\mathfrak{E} = -\,\text{grad}\,\phi = \frac{1}{e}\,\frac{\mathbf{r}}{r}\,\frac{dV}{dr}\,, \tag{60.2}$$

where ϕ is the electrostatic potential produced by the nucleus and the electron cloud, and $V = -e\phi$ is the potential energy of the electron in this field. Substituting the above value of $\mathfrak{E}$ in (60.1), the following expression is found for the spin–orbit interaction:

$$\frac{1}{ec}\,\frac{1}{r}\,\frac{dV}{dr}\,\mu_s(\mathbf{v} \times \mathbf{r}) = \frac{1}{m_0^2 c^2}\,\frac{1}{r}\,\frac{dV}{dr}\,\mathbf{NS}\,, \tag{60.3}$$

where $\mathbf{N} = \mathbf{r} \times m_0\mathbf{v}$ is the orbital angular momentum and $\mathbf{S}$ is the spin. If the movement of the electron in the electrostatic field ϕ is described by the correct relativistic equations, then, according to the investigations of Thomas, a factor of $\frac{1}{2}$ has still to be added to the expression (60.3). Thus the final form of the spin–orbit interaction is

$$\mathcal{H} = \frac{1}{2m_0^2 c^2}\,\frac{1}{r}\,\frac{dV}{dr}\,\mathbf{NS}\,. \tag{60.4}$$

This energy is usually very small compared with the electrostatic energy V of the electron, i.e. it can be regarded as a perturbation.

195

It is well known that the electron states in the atom are generally degenerate: there are $2(2l + 1)$ electron states in a shell characterized by the principal quantum number n and the orbital quantum number l. Two possibilities were introduced for distinguishing them. The states of the shell can be described by either the magnetic and spin quantum numbers, m and m_s, or by the total angular momentum quantum number j and its z-component. In the first case the states of the (n, l) shell are given by the functions

$$\psi_{nlmm_s} = R_{nl}(r) Y_l^m(\vartheta, \phi) \chi_{m_s}(s) ,$$

(60.5)

$$m = 0, \pm 1, \pm 2, \ldots, \pm l \quad \text{and} \quad m_s = \pm \frac{1}{2}$$

in the second case the eigenfunctions will be as follows:

$$\psi_{nljm_j} = R_{nl}(r) F_{ljm_j}(\vartheta, \phi, s) ,$$

(60.6)

$$j = l \pm \frac{1}{2} \quad \text{and} \quad m_j = \pm \frac{1}{2}, \pm \frac{3}{2}, \ldots, \pm j .$$

These two sets of functions are totally equivalent; it was shown in Section 38 that they can be obtained from each other by linear combinations. For the investigation of a given problem always the more convenient one may be chosen. At the present moment it will be advantageous to start from the eigenfunctions (60.6) of the total angular momentum as these are at the same time eigenfunctions of the **NS** operator occurring in the expression of the spin–orbit interaction:

$$(\text{NS}) R_{nl} F_{ljm_j} = \frac{1}{2} [J^2 - N^2 - S^2] R_{nl} F_{l, m_j} =$$

$$= \frac{h}{8\pi^2} \left[j(j + 1) - l(l + 1) - \frac{3}{4} \right] R_{nl} F_{ljm_j} .$$

This can be expressed by the Landé g_L factor (39.6):

$$(\text{NS}) R_{nl} F_{ljm_j} = \frac{h^2}{16\pi^2} \frac{2 - g_L}{g_L - 1} R_{nl} F_{ljm_j} .$$

(60.7)

Let us now find the matrix elements of the perturbation $\mathcal{H}$ according to rule (56.23). Due to (60.7) the following result is achieved:

$$\mathcal{H}^{(nl)}_{jm_j, j' m'_j} = \frac{1}{2} \frac{2 - g_L}{g_L - 1} \zeta_{nl} \delta_{jj'} \delta_{m_j m'_j} ,$$

(60.8)

where the orthogonality of the functions F_{ljm_j} of different indexes has been made use of. ζ_{nl} denotes the following integral:

$$\zeta_{nl} = \frac{h^2}{16\pi^2 m_0^2 c^2} \int_0^\infty \frac{1}{r} \frac{dV}{dr} R_{nl}^2(r)\, r^2 dr .$$ (60.9)

Equation (60.8) shows that the perturbation matrix is diagonal: the matrix-element will be different from zero only for $j = j'$ and $m_j = m$. In these cases the diagonal elements are at the same time the solutions of the secular equation as well:

$$w = \frac{1}{2} \frac{2 - g_L}{g_L - 1} \zeta_{nl} .$$ (60.10)

Since g_L depends on the total angular momentum quantum number j, two different energy corrections w are obtained for $j = l + \frac{1}{2}$ and $j = l - \frac{1}{2}$.

On the other hand, g_L is independent of quantum number m_j, and therefore the value w remains $2j + 1$ times degenerate. If expression g_L is substituted in (60.10), then the end result is

$$w = l\,\zeta_{nl}, \qquad \text{if } j = l + \frac{1}{2} \quad \text{a } (2l + 2)\text{-fold root}$$

and (60.11)

$$w = -(l + 1)\,\zeta_{nl}, \quad \text{if } j = l - \frac{1}{2} \quad \text{a } (2l)\text{-fold root.}$$

Summarizing it can be stated that due to the spin–orbit interaction the term originally $2(2l + 1)$ times degenerate splits into two levels, distinguished by the total angular momentum quantum number j. Both levels are $2j + 1$ times degenerate, the mean value of the levels weighted according the multiplicity is

$$\frac{(2l + 2)\, l\zeta_{nl} - 2l(l + 1)\, \zeta_{nl}}{2(2l + 1)} = 0 ,$$

thus the "centre of mass" of the levels coincides with the unperturbed value. It follows from (60.9) that in the case of an attractive potential ξ_{nl} will be positive, in this case the levels corresponding to $j = l - \frac{1}{2}$ lie deeper than those corresponding to $j = l + \frac{1}{2}$.

61. The Quantum Theory of the Zeeman Effect

The Zeeman level splitting has already been introduced in Section 12; now its quantum-mechanical interpretation will be presented. The magnetic energy of an electron in an external magnetic field $\mathfrak{H}$ is $-\mu \cdot \mathfrak{H}$, where μ is the total magnetic moment of the electron. This is composed of the orbital and spin angular momenta. Let us put the z-axis of our coordinate system in the direction of the magnetic field; the magnetic energy in this case is

$$\mathcal{H} = -\mathfrak{H}\mu_z = \frac{e\mathfrak{H}}{2m_0c}(N_z + 2S_z), \qquad (61.1)$$

where $\mathfrak{H}$ is the magnetic field strength. This magnetic energy of the electrons will be regarded as a perturbation compared to the total energy of the atom.

It will be convenient to describe the states n, l of the electron shell by the functions (60.5). These are the eigenfunctions of the perturbations (61.1), therefore the matrix $\mathcal{H}^{(nl)}_{mms, m' m'_s}$, corresponding to (56.23) is diagonal. The diagonal elements, giving at the same time the solutions of the secular equation, are equal to the eigenvalues of the perturbation:

$$w = \frac{e\mathfrak{H}}{2m_0c} \frac{h}{2\pi}(m + 2m_s) = \mu_B\mathfrak{H}(m + 2m_s),$$

$$\qquad (61.2)$$

$$m = 0, \pm 1, \pm 2, \ldots, \pm l \quad \text{and} \quad m_s = \pm \frac{1}{2}.$$

The energy correction depends on the magnetic (m) and spin (m_s) quantum number; this implies that (61.3) accounts for the splitting of the term levels. As $m + 2m_s$ can only be an integer, the splitting takes place at equal intervals of magnitude $\mu_B\mathfrak{H}$. The splitting is complete only for the s terms with an orbital quantum number $l = 0$: this doublet term splits into two parts, according to the two possible values of m_s (see eq. (61.2)). In case of the terms, where $l > 0$, the Zeeman splitting resolves the degeneracy only partly. If all the possible values of m and m_s are written into expression (61.2), then only $2l + 3$ different w values are obtained: $w = 0, \pm\mu_B\mathfrak{H}$, $\pm 2\mu_B\mathfrak{H}, \ldots, \pm(l + 1)\mu_B\mathfrak{H}$. The splitting of the s and p terms are shown in Fig. 31.

The Zeeman effect takes place in the above discussed manner only in the case of relatively strong magnetic fields ($\mathfrak{H} \geq 5 \times 10^4$ gauss). It is called the normal Zeeman effect. In weaker magnetic fields the so-called *anomalous Zeeman effect* can be observed; here the splitting of the terms is much more complicated than shown by (61.2). The explanation of this fact is that if the magnetic field is not too strong then the spin–orbit interaction cannot be neglected besides the magnetic energy of (61.1). Thus the anomalous Zeeman effect is produced by the joint action of perturbations (60.4) and (61.1):

$$\mathcal{H} = \frac{1}{2m_0^2 c^2}\,\frac{1}{r}\,\frac{dV}{dr}\,\mathbf{NS} + \frac{e\mathfrak{H}}{2m_0 c}\,(N_z + 2S_z).\qquad (61.3)$$

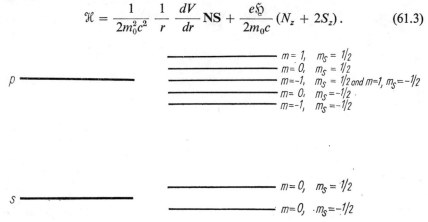

FIG. 31. The Zeeman splitting of the s and p terms

Let us investigate as an example the anomalous Zeeman splitting of the p terms. For $l = 1$ the degree of degeneracy is $2(2 + 1) = 6$, therefore a determinant of order 6 will stand in the secular equation. The elements of this determinant have to be produced from the perturbation $\mathcal{H}$ according to (56.23). The functions ψ^0 can be taken either from (60.5) or from (60.6). Not even practical reasons specify now one system of functions or the other. Let us start with functions (60.5). The effect of spin–orbit interaction operator $\mathbf{NS}$ on the functions $Y_1^m \chi_{ms}$ can be determined from (38.7) and from Table 1 (see p. 77). After elementary calculations the form of the secular equation is

$$\begin{vmatrix}
\zeta + 2\mu_B\mathfrak{H} - w & 0 & 0 & 0 & 0 & 0 \\
0 & -\zeta - w & \sqrt{2}\,\zeta & 0 & 0 & 0 \\
0 & \sqrt{2}\,\zeta & \mu_B\mathfrak{H} - w & 0 & 0 & 0 \\
0 & 0 & 0 & -\mu_B\mathfrak{H} - w & \sqrt{2}\,\zeta & 0 \\
0 & 0 & 0 & \sqrt{2}\,\zeta & -\zeta - w & 0 \\
0 & 0 & 0 & 0 & 0 & \zeta - 2\mu_B\mathfrak{H} - w
\end{vmatrix} = 0,$$

where ζ denotes the integral (60.9). By solving the secular equation six different w values are obtained, thus the degeneracy is completely resolved.

14*

The solutions are the following:

$$w = \zeta + 2\mu_B\mathfrak{H}$$

$$w = \frac{1}{2}\left[-\zeta + \mu_B\mathfrak{H} + \sqrt{9\zeta^2 + 2\mu_B\mathfrak{H}\zeta + \mu_B^2\mathfrak{H}^2}\right]$$

$$w = \frac{1}{2}\left[-\zeta + \mu_B\mathfrak{H} - \sqrt{9\zeta^2 + 2\mu_B\mathfrak{H}\zeta + \mu_B^2\mathfrak{H}^2}\right]$$

$$(61.4)$$

$$w = \frac{1}{2}\left[-\zeta - \mu_B\mathfrak{H} + \sqrt{9\zeta^2 - 2\mu_B\mathfrak{H}\zeta + \mu_B^2\mathfrak{H}^2}\right]$$

$$w = \frac{1}{2}\left[-\zeta - \mu_B\mathfrak{H} - \sqrt{9\zeta^2 - 2\mu_B\mathfrak{H}\zeta + \mu_B^2\mathfrak{H}^2}\right]$$

$$w = \zeta - 2\mu_B\mathfrak{H}.$$

The energy corrections as functions of the magnetic field are shown in Fig. 32. If $\mathfrak{H} = 0$, i.e. there is no external magnetic field, we have a splitting:

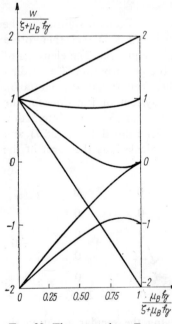

$$w = \zeta \quad \text{(four-fold root)}$$
and
$$w = -2\zeta \quad \text{(two-fold root)},$$

corresponding to the spin–orbit interaction, and for $\mu_B\mathfrak{H} \gg \zeta$ expression (61.4) goes over into the values describing the normal Zeeman effect:

$$w = 2\mu_B\mathfrak{H}, \quad w = \mu_B\mathfrak{H},$$

$$w = 0 \quad \text{(two-fold root)},$$

$$w = -\mu_B\mathfrak{H}, \quad w = -2\mu_B\mathfrak{H}.$$

Fig. 32. The anomalous Zeeman splitting of the p terms

To be able to determine the strength of the magnetic field when the normal Zeeman effect takes place, let us calculate the ζ value for the $2p$ term of H. The potential V originates in H only from the nucleus: $V = -e^2/r$, the

200

radial wave function can be found from (44.22):

$$R_{2p}(r) = (24a_H^5)^{-\frac{1}{2}} r e^{-r/2a_H} .$$

Substituting these in (60.9), we find

$$\zeta = \frac{2\pi^4 e^8 m_0}{3h^4 c^2} = 6.04 \times 10^{-5} \, \text{eV} .$$

The normal Zeeman effect takes place if $\mu_B \mathfrak{H} \gg \zeta$, thus if

$$\mathfrak{H} \gg \frac{\zeta}{\mu_B} = 1.04 \times 10^4 \, \text{gauss} .$$

62. The Stark Effect in Hydrogen

The quantum-mechanical perturbation calculus accounts also for the Stark effect described in Section 12. Here only the simplest case, the Stark effect of the terms of the hydrogen-like atoms, will be dealt with. The Schrödinger equation of the unperturbed atom has already been solved in Section 44. If the z-axis of the coordinate system is taken in the direction of the external electric field, the perturbation energy will be

$$\mathcal{H} = -e\mathfrak{E}z = -e\mathfrak{E}r \cos \vartheta , \tag{62.1}$$

where $\mathfrak{E}$ is the electric field strength. As neither the energy of the unperturbed atom nor the perturbation contains quantities related to the electron spin, for the Stark effect the spin can be neglected.

A single eigenfunction corresponds to the ground state of the unperturbed atom (principal quantum number $n = 1$):

$$\psi_{100} = R_{10}(r) Y_0^0(\vartheta, \phi) = \left(\frac{Z^3}{\pi a_H^3} \right)^{\frac{1}{2}} e^{-Zr/a_H} .$$

The perturbation of the ground state is given in first approximation by

$$w = \int \psi_{100}^* \mathcal{H} \psi_{100} \, dv = 0 , \tag{62.2}$$

because the ψ eigenfunction is spherically symmetrical, and the perturbation $\mathcal{H}$ is an odd function. Let us now deal with the first excited state $n = 2$. The energy of this is $E_2 = -e^2/8a_H$, and four eigenfunctions correspond to this energy:

$$\psi_{200} = R_{20}(r)\, Y_0^0(\vartheta, \phi) = \left(\frac{Z^3}{8\pi a_H^3}\right)^{\frac{1}{2}} \left(1 - \frac{Zr}{2a_H}\right) e^{-Zr/2a_H},$$

$$\psi_{211} = R_{21}(r)\, Y_1^1(\vartheta, \phi) = \left(\frac{Z^3}{8\pi a_H^3}\right)^{\frac{1}{2}} \frac{Zr}{2a_H} e^{-Zr/2a_H} \sin\vartheta \, \frac{e^{i\phi}}{\sqrt{2}},$$

$$\psi_{210} = R_{21}(r)\, Y_1^0(\vartheta, \phi) = \left(\frac{Z^3}{8\pi a_H^3}\right)^{\frac{1}{2}} \frac{Zr}{2a_H} e^{-Zr/2a_H} \cos\vartheta,$$

$$\psi_{21,-1} = R_{21}(r)\, Y_1^{-1}(\vartheta, \phi) = \left(\frac{Z^3}{8\pi a_H^3}\right)^{\frac{1}{2}} \frac{Zr}{2a_H} e^{-Zr/2a_H} \sin\vartheta \, \frac{e^{-i\phi}}{\sqrt{2}}.$$

Let us form the matrix elements $\mathcal{H}_{lm,\,l'm'}^{(2)}$ of the perturbation by these eigenfunctions according to rule (56.23). Only two of the matrix elements, namely

$$\mathcal{H}_{10,00}^{(2)} = \mathcal{H}_{00,10}^{(2)} = \int \psi_{210}^* \mathcal{H}\psi_{200}\, dv \tag{62.3}$$

will differ from zero, all the others have odd functions as integrand. Integral (62.3) can be computed using elementary methods, the result is $-3e\mathfrak{E}a_H$. Thus we obtain the following secular equation for the energy correction w:

$$\begin{vmatrix} -w & -3e\mathfrak{E}a_H & 0 & 0 \\ -3e\mathfrak{E}a_H & -w & 0 & 0 \\ 0 & 0 & -w & 0 \\ 0 & 0 & 0 & -w \end{vmatrix} = 0. \tag{62.4}$$

Or in expanded form

$$w^2(w^2 - 9e^2\mathfrak{E}^2 a_H^2) = 0. \tag{62.5}$$

The solutions of this algebraic equation of fourth degree are

$$\begin{aligned} w &= 3e\mathfrak{E}a_H, \\ w &= 0 \quad \text{(double root)}, \\ w &= -3e\mathfrak{E}a_H. \end{aligned} \tag{62.6}$$

Thus the energy term for $n = 2$ splits in an electric field into three parts, the energy of these are

$$-\frac{e^2}{8a_H} + 3e\mathfrak{E}a_H, \qquad -\frac{e^2}{8a_H}, \qquad -\frac{e^2}{8a_H} - 3e\mathfrak{E}a_H.$$

The end result is: in an electric field instead of the spectrum line corresponding to the transition $n = 2 \to n = 1$ three lines are found. This is backed also by the experimental results. We only mention that the level splitting due to the Stark effect is very small compared to the unperturbed energy. The relative energy correction namely is

$$\eta = \frac{3e\mathfrak{E}a_H}{e^2/8a_H} = 0.466 \times 10^{-8}\mathfrak{E},$$

if $\mathfrak{E}$ is measured in V cm^{-1}. This does not even reach one ten-thousandth for a field strength of $\mathfrak{E} = 2 \times 10^4$ V cm^{-1}.

63. Two-electron Problems

Let us regard a conservative potential field with a potential $V(\mathbf{r})$. The energy operator of the particle m_0 moving in this potential field is

$$H = -\frac{h^2}{8\pi^2 m_0}\nabla^2 + V(\mathbf{r}) . \tag{63.1}$$

In the following it is supposed that the Schrödinger equation belonging to this operator has already been solved, thus the eigenvalues E_k and eigenfunctions $\psi_k(\mathbf{r})$ are known. Now we will deal with the problem when two identical particles move in the potential field $V(\mathbf{r})$. A He-atom or a H$_2$ molecule is, for instance, such a system. The energy operator of the two-particle system is

$$\mathsf{H} = -\frac{h^2}{8\pi^2 m_0}\nabla_1^2 - \frac{h^2}{8\pi^2 m_0}\nabla_2^2 + V(\mathbf{r}_1) + V(\mathbf{r}_2) + v(\mathbf{r}_1, \mathbf{r}_2) , \tag{63.2}$$

where $v(\mathbf{r}_1, \mathbf{r}_2)$ is the interaction energy of the two particles. The H operator can be written in the following form:

$$\mathsf{H} = H^0 + \mathfrak{H} , \tag{63.3}$$

$$H^0 = H(1) + H(2) , \tag{63.4}$$

$$\mathfrak{H} = v(\mathbf{r}_1, \mathbf{r}_2) , \tag{63.5}$$

where $H(1)$ and $H(2)$ denote the energy expressions (63.1) of particle 1 and particle 2, respectively. The eigenvalues and eigenfunctions of operator H will be determined by perturbation calculus; the interaction of the two

203

particles is regarded as the perturbation. First of all the solution of the Schrödinger equation corresponding to the operator H^0 has to be known:

$$H^0\psi^0 = E^0\psi^0. \tag{63.6}$$

It is easy to prove that the function

$$\psi_{kl}^0(\mathbf{r}_1, \mathbf{r}_2) = \psi_k(\mathbf{r}_1)\,\psi_l(\mathbf{r}_2) \tag{63.7}$$

is a solution of the Schrödinger equation:

$$H^0\psi_{kl}^0(\mathbf{r}_1, \mathbf{r}_2) = \psi_l(\mathbf{r}_2)\,H(1)\,\psi_k(\mathbf{r}_1) + \psi_k(\mathbf{r}_1)\,H(2)\,\psi_l(\mathbf{r}_2) =$$
$$= (E_k + E_l)\,\psi_k(\mathbf{r}_1)\,\psi_l(\mathbf{r}_2) = (E_k + E_l)\,\psi_{kl}^0(\mathbf{r}_1, \mathbf{r}_2),$$

and the corresponding eigenvalue is

$$E_{kl}^0 = E_k + E_l. \tag{63.8}$$

This eigenvalue is degenerate, because not only the eigenfunction of (63.7), but also the following eigenfunction corresponds to it:

$$\psi_{lk}^0(\mathbf{r}_1, \mathbf{r}_2) = \psi_l(\mathbf{r}_1)\,\psi_k(\mathbf{r}_2). \tag{63.9}$$

Only the $k = l$ case is an exception, because then (63.7) and (63.9) are equal. That means that only a single eigenfunction

$$\psi_{kk}^0(\mathbf{r}_1, \mathbf{r}_2) = \psi_k(\mathbf{r}_1)\,\psi_k(\mathbf{r}_2) \tag{63.10}$$

corresponds to the eigenvalue $E_{kk}^0 = 2E_k$.

Let us now come to the determination of the energy correction w originating from the interaction. First the $k = l$ case is considered, when expression (56.16) valid for the simple eigenvalues is applicable; the integration, of course, has to be extended over every variable occurring in the wave function:

$$w = \int \psi_{kk}^{0*}\,\mathcal{H}\,\psi_{kk}^0\,dv = \int\int \psi_k^*(\mathbf{r}_1)\,\psi_k^*(\mathbf{r}_2)\,v(\mathbf{r}_1, \mathbf{r}_2)\,\psi_k(\mathbf{r}_1)\,\psi_k(\mathbf{r}_2)\,dv_1 dv_2. \tag{63.11}$$

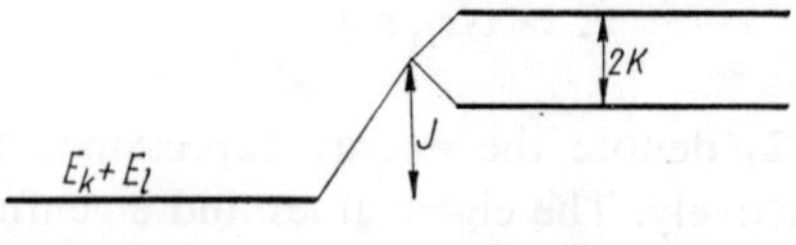

FIG. 33. The energy perturbation due to the interaction between the two electrons

204

Let us now regard the $k \neq l$ case. Now we have two eigenfunctions, ψ_{kl}^0 and ψ_{lk}^0; according to this the energy correction w is found from a secular equation of order 2:

$$\begin{vmatrix} J_1 - w & K_1 \\ K_2 & J_2 - w \end{vmatrix} = 0 . \tag{63.12}$$

The expressions of the matrix elements are

$$J_1 = \int \psi_{kl}^{0*} \mathcal{H} \psi_{kl}^0 dv = \int \int \psi_k^*(\mathbf{r}_1)\, \psi_l^*(\mathbf{r}_2)\, v(\mathbf{r}_1, \mathbf{r}_2)\, \psi_k(\mathbf{r}_1)\, \psi_l(\mathbf{r}_2)\, dv_1 dv_2 ,$$

$$K_1 = \int \psi_{kl}^{0*} \mathcal{H} \psi_{lk}^0 dv = \int \int \psi_k^*(\mathbf{r}_1)\, \psi_l^*(\mathbf{r}_2)\, v(\mathbf{r}_1, \mathbf{r}_2)\, \psi_l(\mathbf{r}_1)\, \psi_k(\mathbf{r}_2)\, dv_1 dv_2 ,$$

$$K_2 = \int \psi_{lk}^{0*} \mathcal{H} \psi_{kl}^0 dv = \int \int \psi_l^*(\mathbf{r}_1)\, \psi_k^*(\mathbf{r}_2)\, v(\mathbf{r}_1, \mathbf{r}_2)\, \psi_k(\mathbf{r}_1)\, \psi_l(\mathbf{r}_2)\, dv_1 dv_2 ,$$

$$J_2 = \int \psi_{lk}^{0*} \mathcal{H} \psi_{lk}^0 dv = \int \int \psi_l^*(\mathbf{r}_1)\, \psi_k^*(\mathbf{r}_2)\, v(\mathbf{r}_1, \mathbf{r}_2)\, \psi_l(\mathbf{r}_1)\, \psi_k(\mathbf{r}_2)\, dv_1 dv_2 . \tag{63.13}$$

The interaction energy of identical, indistinguishable particles does not change if the two particles are reversed

$$v(\mathbf{r}_1, \mathbf{r}_2) = v(\mathbf{r}_2, \mathbf{r}_1) .$$

This character of the interaction has a very important consequence. If in the first two expressions of (63.13) the substitution $\mathbf{r}_1 \rightleftharpoons \mathbf{r}_2$ is made and the above symmetry of v is taken into consideration, then we see that $J_1 = J_2$ and $K_1 = K_2$. Thus the form of the secular equation is

$$\begin{vmatrix} J - w & K \\ K & J - w \end{vmatrix} = 0 . \tag{63.14}$$

J is often called the normal and K the exchange interaction. Expanding the above determinant the following equation of second degree is obtained:

$$w^2 - 2Jw + J^2 - K^2 = 0 .$$

Its solutions are

$$w = J \pm K . \tag{63.15}$$

And thus the eigenvalue of the energy operator H is

$$E = E_k + E_l + J \pm K , \tag{63.16}$$

which is the sum of the unperturbed energies $E_k + E_l$ and the normal energy J with the exchange energy K added either with a $+$ or a $-$ sign.

Due to the interaction the energy level is displaced by J and splits into two parts; the magnitude of the splitting is $2K$.

Let us determine the eigenfunctions corresponding to the eigenvalues (63.16). In zeroth approximation, according to (56.27)

$$\psi^0 = c_1\psi^0_{kl} + c_2\psi^0_{lk}, \tag{63.17}$$

where the coefficients c_1 and c_2 satisfy the following homogeneous system of linear simultaneous equations

$$(J - w)\,c_1 + Kc_2 = 0\,,$$
$$Kc_1 + (J - w)\,c_2 = 0\,. \tag{63.18}$$

Let us take first the root $w = J + K$. Now both equations of (63.18) show that $c_1 = c_2$. Their common value is determined by the condition of normalization of the eigenfunction (63.17). From this the value $\sqrt{\tfrac{1}{2}}$ is found. Thus the normalized eigenfunction is

$$\psi^0(\mathbf{r}_1, \mathbf{r}_2) = \frac{1}{\sqrt{2}}\left[\psi_k(\mathbf{r}_1)\,\psi_l(\mathbf{r}_2) + \psi_l(\mathbf{r}_1)\,\psi_k(\mathbf{r}_2)\right]. \tag{63.19}$$

This eigenfunction is a symmetrical function of the coordinates of the two particles:

$$\psi^0(\mathbf{r}_1, \mathbf{r}_2) = \psi^0(\mathbf{r}_2, \mathbf{r}_1)\,.$$

The other solution of the secular equation is $w = J - K$. In this case both equations of (63.18) yield the result that $c_1 = -c_2$. In accordance with this the normalized eigenfunction is

$$\psi^0(\mathbf{r}_1, \mathbf{r}_2) = \frac{1}{\sqrt{2}}\left[\psi_k(\mathbf{r}_1)\,\psi_l(\mathbf{r}_2) - \psi_l(\mathbf{r}_1)\,\psi_k(\mathbf{r}_2)\right]. \tag{63.20}$$

This eigenfunction is an antisymmetrical function of the coordinates.

$$\psi^0(\mathbf{r}_1, \mathbf{r}_2) = -\psi^0(\mathbf{r}_2, \mathbf{r}_1)\,.$$

Let us apply our result to the He atom and He-like ions, where two electrons revolve around a nucleus with a charge of Ze. The one-particle operator of (63.1) will be now

$$H = -\frac{h^2}{8\pi^2 m_0}\nabla^2 - \frac{Ze^2}{r}\,,$$

which is the energy expression of the H-like ions. The eigenvalues and eigen-functions of this operator have been determined in Section 44.

First the ground state of He should be investigated. In this case both electrons are in the $1s$ state and thus the eigenvalue of the interaction-free operator is

$$E_{11}^0 = 2E_1 = - \frac{4\pi^2 m_0 Z^2 e^4}{h^2} = - \frac{Z^2 e^2}{a_H}, \qquad (63.21)$$

This is twice the value of the ionization energy of H. The corresponding unperturbed eigenfunction (63.10) is

$$\psi^0(\mathbf{r}_1, \mathbf{r}_2) = \psi_{100}(\mathbf{r}_1)\,\psi_{100}(\mathbf{r}_2) = \frac{Z^3}{\pi a_H^3}\, e^{-Z/a_H\,(r_1 + r_2)}. \qquad (63.22)$$

Let us now determine the energy correction (63.11) produced by the inter-action. v denotes now the Coulomb interaction of the two electrons: $v(r_1, r_2) = e^2/r_{12}$, where $r_{12} = |\,\mathbf{r}_1 - \mathbf{r}_2\,|$

$$w = \frac{Z^6 e^2}{\pi^2 a_H^6} \int\!\!\int \frac{e^{-(2Z/a_H)\,(r_1 + r_2)}}{r_{12}}\, dv_1 dv_2 . \qquad (63.23)$$

According to a well-known result of the theory of electrostatic potentials for every spherically symmetrical function $\rho(r)$, we have

$$\int \frac{\rho(r_2)}{r_{12}}\, dv_2 = \frac{1}{r_1} \int\limits_0^{r_1} \rho(r_2)\, 4\pi r_2^2 dr_2 + \int\limits_{r_1}^{\infty} \rho(r_2)\, 4\pi r_2 dr_2, \qquad (63.24)$$

therefore

$$\int \frac{e^{-(2Z/a_H)r_2}}{r_{12}}\, dv_2 = \frac{\pi a_H^2}{Z^2} \left\{ \frac{a_H}{Zr_1} - \left(1 + \frac{a_H}{Zr_1} \right) e^{-2Zr_1/a_H} \right\}.$$

This enables the computation of the double integral of (63.23)

$$w = \frac{Z^6 e^2}{\pi^2 a_H^6} \cdot \frac{\pi a_H^2}{Z^2} \int\limits_0^{\infty} \left\{ \frac{a_H}{Zr_1} - \left(1 + \frac{a_H}{Zr_1} \right) e^{-2Zr_1/a_H} \right\} e^{-2Zr_1/a_H}\, 4\pi r_1^2 dr_1 = \frac{5}{8} \frac{Ze^2}{a_H}. \qquad (63.25)$$

Thus the ground-state energy of the He atom and of He-like ions is given by the following expression:

$$E_{11} = - \left(Z^2 - \frac{5}{8} Z \right) \frac{e^2}{a_H}. \qquad (63.26)$$

The ionization energy is the work to be performed during the detachment of one of the electrons. After ionization, an H-like ion with one electron is left behind, the energy of this is $E_1 = -Z^2e^2/2a_H$. And thus the ionization work is

$$I = E_1 - E_{11} = \left(\frac{1}{2}Z^2 - \frac{5}{8}Z\right)\frac{e^2}{a_H}. \tag{63.27}$$

The universal constant occurring here, e^2/a_H, is the most natural unit of energy in theoretical nuclear and molecular physics. Its value is $e^2/a_H = 27.23$ eV.

In the first excited state, following the ground state, one of the electrons gets in a $2s$ level, the other stays in the $1s$ level. The corresponding eigenvalue of the interaction-free operator H^0 now is

$$E_{12}^0 = E_1 + E_2 = -\frac{Z^2e^2}{2a_H}\left(\frac{1}{1^2} + \frac{1}{2^2}\right) = -\frac{5Z^2e^2}{8a_H}. \tag{63.28}$$

According to (63.15) the energy correction produced by the interaction builds up from the normal (J) and exchange (K) energies. The values of these energies are given by integrals (63.13), only in the place of ψ_k and ψ_l the eigenfunctions corresponding to the $1s$ and $2s$ states of H have to be substituted.

$$J = \int\int \psi_{100}^*(\mathbf{r}_1)\,\psi_{200}^*(\mathbf{r}_2)\,\frac{e^2}{r_{12}}\,\psi_{100}(\mathbf{r}_1)\,\psi_{200}(\mathbf{r}_2)\,dv_1dv_2\,,$$

$$K = \int\int \psi_{100}^*(\mathbf{r}_1)\,\psi_{200}^*(\mathbf{r}_2)\,\frac{e^2}{r_{12}}\,\psi_{200}(\mathbf{r}_1)\,\psi_{100}(\mathbf{r}_2)\,dv_1dv_2\,.$$

If the expressions of the H eigenfunctions

$$\psi_{100}(\mathbf{r}) = \left(\frac{Z^3}{\pi a_H^3}\right)^{\frac{1}{2}} e^{-Zr/a_H},$$

$$\psi_{200}(\mathbf{r}) = \left(\frac{Z^3}{8\pi a_H^3}\right)^{\frac{1}{2}}\left(1 - \frac{Zr}{2a_H}\right) e^{-Zr/2a_H},$$

are substituted in these expressions, then integrals are found that can be solved easily by the help of the result (63.24). The results are

$$J = \frac{17}{81}\frac{Ze^2}{a_H} \quad \text{and} \quad K = \frac{16}{729}\frac{Ze^2}{a_H}. \tag{63.29}$$

The first excited energy of the He atom and of the He-like ions thus is

$$E_{12} = E_{12}^0 + J \pm K = \left(-\frac{5}{8}Z^2 + \frac{17}{81}Z \pm \frac{16}{729}Z\right)\frac{e^2}{a_H}. \quad (63.30)$$

The upper sign corresponds to the symmetric, the lower one to the anti-symmetric eigenfunction. The difference between (63.26) and (63.29) yields the energy necessary for the excitation. For the neutral He atom the following results are obtained:

	Symmetric state	Antisymmetric state
$\Delta E = E_{12} - E_{11}$:	$0.7135\,\dfrac{e^2}{a_H}$	$0.6257\,\dfrac{e^2}{a_H}$.
Experimental value :	$0.7571\,\dfrac{e^2}{a_H}$	$0.7279\,\dfrac{e^2}{a_H}$.

A better agreement cannot be expected from the first approximation of the perturbation calculus, as the interaction e^2/r_{12} of the two electrons is not smaller in order of magnitude than the unperturbed energy.

Up to now only the movement of the two electrons in space has been investigated, let us now deal with the spin states of the two electrons. Two electrons can occupy the two possible spin states in four different forms:

$$\chi_{++}(s_1, s_2) = \chi_{+\frac{1}{2}}(s_1)\,\chi_{+\frac{1}{2}}(s_2)\,,$$
$$\chi_{+-}(s_1, s_2) = \chi_{+\frac{1}{2}}(s_1)\,\chi_{-\frac{1}{2}}(s_2)\,,$$
$$\chi_{-+}(s_1, s_2) = \chi_{-\frac{1}{2}}(s_1)\,\chi_{+\frac{1}{2}}(s_2)\,, \quad (63.31)$$
$$\chi_{--}(s_1, s_2) = \chi_{-\frac{1}{2}}(s_1)\,\chi_{-\frac{1}{2}}(s_2)\,.$$

It is often convenient to apply instead of these functions the symmetric and antisymmetric combinations produced from them:

$$\chi_{11}(s_1, s_2) = \chi_{+\frac{1}{2}}(s_1)\,\chi_{+\frac{1}{2}}(s_2)\,,$$
$$\chi_{10}(s_1, s_2) = \frac{1}{\sqrt{2}}\left[\chi_{+\frac{1}{2}}(s_1)\,\chi_{-\frac{1}{2}}(s_2) + \chi_{-\frac{1}{2}}(s_1)\,\chi_{+\frac{1}{2}}(s_2)\right],$$
$$\quad (63.32)$$
$$\chi_{1,-1}(s_1, s_2) = \chi_{-\frac{1}{2}}(s_1)\,\chi_{-\frac{1}{2}}(s_2)\,,$$
$$\chi_{00}(s_1, s_2) = \frac{1}{\sqrt{2}}\left[\chi_{+\frac{1}{2}}(s_1)\,\chi_{-\frac{1}{2}}(s_2) - \chi_{-\frac{1}{2}}(s_1)\,\chi_{+\frac{1}{2}}(s_2)\right].$$

The first three spin functions are symmetric combination of s_1 and s_2, whereas the fourth one is the antisymmetric combination. It is obvious that these functions are the eigenfunctions of the resultant spin $\mathbf{S} = \mathbf{S}_1 + \mathbf{S}_2$ of the two electrons. According to the formulae gathered in Table 1 (see p. 77):

$$(\mathbf{S}_1 + \mathbf{S}_2)^2 \chi_{\mathscr{S}\mathscr{M}} = \frac{\mathscr{S}(\mathscr{S} + 1)\,h^2}{4\pi^2} \chi_{\mathscr{S}\mathscr{M}} \,,$$

$$(S_{1z} + S_{2z}) \chi_{\mathscr{S}\mathscr{M}} = \frac{\mathscr{M}h}{2\pi} \chi_{\mathscr{S}\mathscr{M}} \,, \tag{63.33}$$

$$(\mathscr{S} = 0 \quad \text{or} \quad 1 \quad \text{and} \quad \mathscr{M} = -\mathscr{S}, 0, \mathscr{S}) \,.$$

These expressions show that the resultant spin quantizes according to the general rules introduced in Section 38. In (63.32) the symmetric functions belong to the resultant spin value $\mathscr{S} = 1$. They are distinguished by the three possible values of the z-component quantum number $\mathscr{M}$. The antisymmetric combination corresponds to the $\mathscr{S} = \mathscr{M} = 0$ value. In an external magnetic field the magnetic energy of states corresponding to different $\mathscr{M}$ values will be different. Due to this the symmetric states corresponding to $\mathscr{S} = 1$ will split into three different levels (triplet term), the antisymmetric state of $\mathscr{S} = 0$, on the other hand, does not split (singlet term).

The analysis of the Zeeman effect of He shows the interesting conclusion that the ground state of He is always a singlet state, the triplet spin combination is never materialized in a He atom ground state. Those excited states, where the $\psi(\mathbf{r}_1, \mathbf{r}_2)$ functions, describing the space movement, are symmetric in $\mathbf{r}_1$ and $\mathbf{r}_2$, are also always singlet terms. To the antisymmetric $\psi(\mathbf{r}_1, \mathbf{r}_2)$ always triplet terms correspond. That means that the states of the He atom can be divided into two large groups. The first group is characterized by symmetric space and antisymmetric spin functions, in the second group the situation is reversed: the eigenfunction is built from an antisymmetric space function and a symmetric spin function. The probability of a transition between the states of the two groups is very small; therefore seemingly two kinds of He are present in nature. The first one, where the terms are singlet, is called *para-helium*, the second one, where the terms are triplet ones, is called *ortho-helium*.

It is characteristic for both groups of terms that if both the space and the spin coordinates of the electrons are interchanged then the eigenfunction behaves antisymmetrically

$$\psi(\mathbf{r}_1 s_1; \mathbf{r}_2 s_2) = -\psi(\mathbf{r}_2 s_2; \mathbf{r}_1 s_1) \,. \tag{63.34}$$

In para-helium the spin-, in ortho-helium the space function changes sign. According to experience this law is valid for the electron cloud of every atom and ion, without any exception; it is valid in general for every system of electrons: the state functions are antisymmetric, i.e. interchanging the coordinates of two electrons leads to a change in sign. This is the general quantum-mechanical formulation of Pauli's principle. The original wording of Pauli, introduced in Section 16, is a simple consequence of this general formulation.

64. The Variational Method

In the quantum theory of atoms and molecules perhaps the most effective method of approximation is the variational method. Here only the determination of the ground state, i.e. of the state with the lowest energy will be dealt with, but the method can be applied for the investigation of the excited states as well.

ϕ should be an arbitrary function of the coordinates of the investigated system. It should only be supposed that it is normalized to unity:

$$\int \phi^* \phi \, dv = 1 . \tag{64.1}$$

Let us expand this arbitrary function ϕ in terms of the eigenfunctions of the energy operator H:

$$\phi = \sum_k c_k \psi_k , \tag{64.2}$$

where ψ_k satisfies the Schrödinger equation belonging to H:

$$H\psi_k = E_k \psi_k . \tag{64.3}$$

As ϕ is a normalized function, the sum of the squares of the c_k coefficients is 1. From (64.2) and (64.1) it follows that

$$\int \phi^* \phi \, dv = \sum_{k,l} c_k^* c_l \int \psi_k^* \psi_l \, dv = 1 .$$

Due to the orthogonality relation (26.8) the integral $\int \psi_k^* \psi_l \, dv = 0$, if $k \neq l$, and for $k = l$ the integral is equal to unity because the ψ_k eigenfunctions are normalized. Therefore

$$\sum_k |c_k|^2 = 1 . \tag{64.4}$$

Let us form the quantum-mechanical mean (expectation) value of the operator H by means of the function ϕ:

$$\int \phi^* H\phi \, dv = \sum_{k,l} c_k^* c_l \int \psi_k^* H\psi_l \, dv = \sum_{k,l} c_k^* c_l E_l \int \psi_k^* \psi_l \, dv = \sum_k |c_k|^2 E_k \,. \quad (64.5)$$

Here the form (64.3) of the Schrödinger equation was used, and the orthogonality and normalization (orthonormal property) of the ψ_k functions has been taken into consideration once again. The right-hand side sum is evidently decreased if instead of the E_k values the smallest eigenvalue, the ground state energy is written

$$\int \phi^* H\phi \, dv = \sum_k |c_k|^2 E_k \geq E_0 \sum_k |c_k|^2 = E_0 \,, \quad (64.6)$$

because, as we can see, the sum of the squares of the coefficients c_k is equal to unity. From the relation $\geq$ the equality can only be realized if the coefficient c_0 of the ground state is equal to unity, and all the other c_k are zero. According to the series expansion of (64.2) this means that $\phi = \psi_0$.

Summarizing our results:

$$\int \phi^* H\phi \, dv \geq E_0 \,, \quad \text{if} \quad \int \phi^* \phi \, dv = 1 \,, \quad (64.7)$$

with the equality holding only if $\phi = \psi_0$. This means that the ground state energy E_0 and eigenfunction ψ_0 can be obtained by solving the following conditional extreme value problem:

$$\int \phi^* H\phi \, dv = \text{minimum} \,, \quad \int \phi^* \phi \, dv = 1 \,. \quad (64.8)$$

An approximate determination of the eigenvalue E_0 and eigenfunction ψ_0 is possible by the Ritz method. A so-called trial function is chosen for ϕ satisfying the condition $\int \phi^* \phi \, dv = 1$, which depends (besides the coordinates) on a number of variable parameters λ. The integral

$$E(\lambda) = \int \phi_\lambda^* H\phi_\lambda \, dv \quad (64.9)$$

is calculated, and the values of the parameter(s) λ are determined so as to make $E(\lambda)$ a minimum. This value should be λ_0. Then approximately

$$E_0 = E(\lambda_0) \quad \text{and} \quad \psi_0 = \phi_{\lambda_0} \,. \quad (64.10)$$

This approximation provides an upper bound for E_0, i.e. it can be never smaller than the exact E_0.

We illustrate the use of the Ritz variational method by applying it to the ground state of the exponential potential well (42.1).

The energy operator (Hamiltonian) is given by

$$H = - \frac{h^2}{8\pi^2 m_0} \nabla^2 - V_0 e^{-r/a}. \tag{64.11}$$

It is reasonable to look for the trial function for the ground state in the following form:

$$\phi_\lambda(\mathbf{r}) = N e^{-\lambda r}. \tag{64.12}$$

The coefficient N is determined from the normalizing condition:

$$\int \phi_\lambda^* \phi_\lambda \, dv = N^2 \int_0^\infty e^{-2\lambda r} 4\pi r^2 \, dr = \frac{N^2 \pi}{\lambda^3} = 1,$$

$$N = \left(\frac{\lambda^3}{\pi} \right)^{\frac{1}{2}}.$$

Therefore the trial function ϕ_λ is

$$\phi_\lambda(\mathbf{r}) = \left(\frac{\lambda^3}{\pi} \right)^{\frac{1}{2}} e^{-\lambda r}. \tag{64.13}$$

The expectation value of H is

$$E(\lambda) = \int \phi_\lambda^* H \phi_\lambda \, dv = \frac{\lambda^3}{\pi} \int_0^\infty e^{-\lambda r} \left\{ - \frac{h^2}{8\pi^2 m_0} \nabla^2 - V_0 e^{-r/a} \right\} e^{-\lambda r} 4\pi r^2 \, dr.$$

Performing the calculations we find that

$$E(\lambda) = \frac{h^2 \lambda^2}{8\pi^2 m_0} - V_0 \left(\frac{2\lambda a}{2\lambda a + 1} \right)^3. \tag{64.14}$$

From the minimum condition $dE/d\lambda = 0$ the following equation is found for the variational parameter

$$\frac{(2\lambda a + 1)^4}{2\lambda a} = \frac{48\pi^2 m_0 a^2}{h^2} V_0. \tag{64.15}$$

This is an algebraic equation of fourth degree for λ. The equation becomes more clearcut if the substitution $x = 2\lambda a$ and the γ^2 notation of (42.5) is introduced:

$$x^4 + 4x^3 + 6x^2 + \left(4 - \frac{3}{2}\gamma^2\right)x + 1 = 0 . \qquad (64.16)$$

In Section 42 it was shown that from the values for deuterium the numerical value of $\gamma = 3.82$ is obtained. In this case the solution of (64.16) is $x_0 = 1.34$, the corresponding λ value is:

$$\lambda_0 = \frac{x_0}{2a} = 0.307 \times 10^{13} \text{ cm}^{-1} .$$

Substituting this in (64.14) the ground level energy is $E(\lambda_0) = -2.18$ MeV. This is somewhat higher than the exact value $E_0 = -2.23$ MeV.

In a frequently used form of the variational method the trial function ϕ is chosen in the form

$$\phi = c_1 f_1 + c_2 f_2 + \ldots + c_p f_p \qquad (64.17)$$

where $f_1, \ldots, f_p$ are predetermined functions and $c_1, \ldots, c_p$ are variable parameters. Neither the normalization nor the orthogonality (orthonormal properties) of the f_α functions is supposed. From the condition of normalization of the function ϕ we find for the coefficients c_α the condition

$$\sum_{\alpha, \beta=1}^{p} S_{\alpha\beta} c_\alpha^* c_\beta = 1 , \qquad (64.18)$$

where

$$S_{\alpha\beta} = \int f_\alpha^* f_\beta \, dv . \qquad (64.19)$$

The expectation value of H for the function ϕ is given by

$$E(c_1, c_2, \ldots, c_p) = \sum_{\alpha, \beta=1}^{p} H_{\alpha\beta} c_\alpha^* c_\beta , \qquad (64.20)$$

where $H_{\alpha\beta}$ denotes the following matrix element:

$$H_{\alpha\beta} = \int f_\alpha^* H f_\beta \, dv . \qquad (64.21)$$

When requiring the minimum of this energy expression the subsidiary condition (64.18) has to be taken into consideration as well.

214

According to the Lagrange method the unconstrained minimum of

$$W(c_1, c_2, \ldots, c_p) = \sum_{\alpha, \beta = 1}^{p} (H_{\alpha\beta} - \varepsilon S_{\alpha\beta}) \, c_\alpha^* c_\beta$$

is required, where ε is a Lagrange multiplier. From the extremal condition $\partial W / \partial c_\alpha^* = 0$ the following algebraic equation is obtained:

$$\sum_{\beta = 1}^{p} (H_{\alpha\beta} - \varepsilon S_{\alpha\beta}) \, c_\beta = 0 \, . \tag{64.22}$$

This is a system of linear homogeneous equations for the coefficients c_β. For these equations to be consistent the determinant D of the system of equations has to be equal to zero. From this condition the Lagrange multiplier can be determined. If ε is known, (64.22) yields the values of the coefficients c_β. In the actual determination of the c_β values (64.18) has to be considered as well.

If the coefficients c_α and ε are known, the minimum value of energy E can be determined. Let us multiply the minimum condition (64.22) by c_α^* and sum it from $\alpha = 1$ till $\alpha = p$:

$$\sum_{\alpha, \beta = 1}^{p} H_{\alpha\beta} c_\alpha^* c_\beta - \varepsilon \sum_{\alpha, \beta = 1}^{p} S_{\alpha\beta} c_\alpha^* c_\beta = 0 \, .$$

According to (64.20) the first sum is equal to the energy E itself, and according to (64.18) the value of the second equation is equal to unity. As an end result we have $E = \varepsilon$, that is, the energy minimum equals the Lagrange multiplier. From

$$D \equiv \begin{vmatrix} H_{11} - ES_{11} & H_{12} - ES_{12} \ldots & H_{1p} - ES_{1p} \\ H_{21} - ES_{21} & H_{22} - ES_{22} \ldots & H_{2p} - ES_{2p} \\ \cdot & \cdot & \cdot \\ \cdot & \cdot & \cdot \\ \cdot & \cdot & \cdot \\ H_{p1} - ES_{p1} & H_{p2} - ES_{p2} \ldots & H_{pp} - ES_{pp} \end{vmatrix} = 0 \tag{64.23}$$

where D is the determinant of (64.22). E can be computed without difficulty. We mention that as (64.23) is an algebraic equation of degree p, it can have several solutions. The energy of the ground state is approximated by the smallest root, the other roots correspond to the energies of the excited states. For these latter ones (64.23) usually is a rather poor approximation.

65. The Ground State of the Helium Atom

The problem of the helium atom and of helium-like ions has been solved in Section 63 by using perturbation calculus. Now the same question will be investigated by means of the variational method. If the trial function ϕ is chosen well, this latter method gives a much higher accuracy than perturbation theory. The Hamiltonian for such a system with nuclear charge Z takes the form (according to the model of (63.2))

$$H = - \frac{h^2}{8\pi^2 m_0} \nabla_1^2 - \frac{h^2}{8\pi^2 m_0} \nabla_2^2 - \frac{Ze^2}{r_1} - \frac{Ze^2}{r_2} + \frac{e^2}{r_{12}}. \quad (65.1)$$

The trial function is chosen in a form similar to (63.22), only the variable parameter Z^* is written instead of the atomic number Z.

$$\phi(\mathbf{r}_1, \mathbf{r}_2) = \frac{Z^*}{\pi a_H^3} e^{-(Z^*/a_H)(r_1+r_2)}. \quad (65.2)$$

The expectation value of the energy operator H is composed of three terms: the kinetic energy of the electrons (E_k), the potential energy (E_c) of the nuclei with a charge Z and the interaction energy (E_w) of the two electrons. By simple calculations the following expressions are found for these energy terms:

$$E_k = - \frac{h^2}{8\pi^2 m_0} \int\int \phi^*(\mathbf{r}_1, \mathbf{r}_2)\, (\nabla_1^2 + \nabla_2^2)\, \phi(\mathbf{r}_1, \mathbf{r}_2)\, dv_1 dv_2 = \frac{Z^{*2}e^2}{a_H},$$

$$E_c = - \int\int \phi^*(\mathbf{r}_1, \mathbf{r}_2) \left(\frac{Ze^2}{r_1} + \frac{Ze^2}{r_2} \right) \phi(\mathbf{r}_1, \mathbf{r}_2)\, dv_1 dv_2 = - \frac{2ZZ^*e^2}{a_H},$$

$$E_w = \int\int \phi^*(\mathbf{r}_1, \mathbf{r}_2) \frac{e^2}{r_{12}} \phi(\mathbf{r}_1, \mathbf{r}_2)\, dv_1 dv_2 = \frac{5}{8} \frac{Z^*e^2}{a_H}.$$

Therefore the total energy $E = E_k + E_c + E_w$ is the following function of Z^*:

$$E(Z^*) = \left(Z^{*2} - 2ZZ^* + \frac{5}{8} Z^* \right) \frac{e^2}{Z}. \quad (65.3)$$

If $Z^* = Z$ is written in this equation the result of perturbation calculus, formula (63.26) is regained. Now we are going to determine Z^* from the energy minimum; this result can only be better than (63.26). The Z^* value corresponding to the energy minimum is called the effective atomic number.

216

The condition of minimum is

$$\frac{dE}{dZ^*} = \left(2Z^* - 2Z + \frac{5}{8}\right)\frac{e^2}{a_H} = 0.$$

From this the effective atomic number is

$$Z^* = Z - \frac{5}{16}. \tag{65.4}$$

Hence we have by substituting in the expression of the energy

$$E = -\left(Z^2 - \frac{5}{8}Z + \frac{25}{256}\right)\frac{e^2}{a_H}. \tag{65.5}$$

Subtracting this E value from the energy $E_1 = Z^2 e^2 / 2a_H$ of the hydrogen-like ion, the ionization potential is obtained

$$I = E_1 - E = \left(\frac{1}{2}Z^2 - \frac{5}{8}Z + \frac{25}{256}\right)\frac{e^2}{a_H}. \tag{65.6}$$

The ionization potential proved to be by $25/256 \; e^2/a_H$ higher than for perturbation theory. Numerical data are collected in Table 7.

TABLE 7. *The Ionization Potential of the Helium Atom and Helium-like Ions*

Z	Name	I from (63.27)	I from (65.6)	I experimental value
2	He	$0.750 \; e^2/a_H$	$0.848 \; e^2/a_H$	$0.9035 \; e^2/a_H$
3	Li$^+$	2.625	2.723	2.7798
4	Be^{++}	5.500	5.598	5.6560
5	B^{3+}	9.375	9.473	9.5320
6	C^{4+}	14.250	14.348	14.4070

Hylleraas performed elaborate calculations for the He atom ($Z = 2$). He used the trial function

$$\phi(\mathbf{r}_1, \mathbf{r}_2) = Ne^{-(Z^*/a_H)(r_1 + r_2)} [1 + c_1 r_{12} + c_2(r_1 - r_2)^2 + c_3(r_1 + r_2)$$
$$+ c_4(r_1 + r_2)^2 + c_5 r_{12}^2 + c_6(r_1 + r_2) r_{12}]$$

217

for the ground state eigenfunction, where $Z^*, c_1, c_2, \ldots, c_6$ are the variable parameters, N is a normalization constant. He found $E = -2.90324 \, e^2/a_H$ as the energy minimum, from this the ionization potential is

$$I = 0.90324 \, e^2/a_H \,,$$

in good agreement with experiment.

66. The H_2^+ Molecular Ion

The H_2^+ molecular ion is the simplest chemically bound system; it consists of a single electron bound to two protons (H nuclei). The resulting electric charge of the entire system is $+e$; according to classical electrodynamics such a system has no stable bound state. The fact that in nature H_2^+ molecular ions can be observed is due to the quantum-mechanical effects. Quantum mechanics can interpret not only the existence of the H_2^+ molecular ion, but gives numerical results for the dissociation energy and equilibrium internuclear distance being in accordance with experimental data.

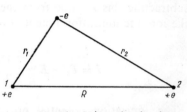

FIG. 34. The H_2^+ molecular ion

If R is the internuclear distance and r_1 and r_2 are the distance of the electron from the nuclei at 1 and 2 respectively (see Fig. 34), we have for the Hamiltonian

$$H = -\frac{h^2}{8\pi^2 m_0} \nabla^2 - \frac{e^2}{r_1} - \frac{e^2}{r_2} + \frac{e^2}{R} \,. \tag{66.1}$$

The ground state will be determined by the variational method. If the internuclear distance R were very large, the electron would obviously form an H-atom with one of the nuclei, and the other bare nucleus would have a vanishingly small influence on this system. Therefore for $R \to \infty$ the wave function will be either

$$\psi_{100}(r_1) = \left(\frac{1}{\pi a_H^3}\right)^{\frac{1}{2}} e^{-r_1/a_H} \,, \tag{66.2}$$

an eigenfunction corresponding to a $1s$ orbit around nucleus 1, or

$$\psi_{100}(r_2) = \left(\frac{1}{\pi a_H^3}\right)^{\frac{1}{2}} e^{-r_2/a_H}, \tag{66.3}$$

an eigenfunction corresponding to a hydrogen orbital at nucleus 2. For finite R distance we therefore take as our trial wave function a linear combination of atomic hydrogen orbitals which can be written in the form

$$\phi(\mathbf{r}) = c_1\psi_{100}(r_1) + c_2\psi_{100}(r_2), \tag{66.4}$$

where c_1 and c_2 are variable parameters. This trial function corresponds to the general expression of (64.17), so that the energy can be determined by using the method discussed at the end of Section 64. As (66.4) contains two terms, the energy matrix (64.21) and the overlap matrix will be of order 2. The expressions of the matrix elements are

$$H_{11} = \int \psi_{100}^*(r_1)\, H\psi_{100}(r_1)\, dv = -\frac{e^2}{2a_H} + \frac{e^2}{R} - P,$$

$$H_{12} = \int \psi_{100}^*(r_1)\, H\psi_{100}(r_2)\, dv = -\frac{e^2}{2a_H} S + \frac{e^2}{R} S - Q,$$

$$\tag{66.5}$$

$$H_{21} = \int \psi_{100}^*(r_2)\, H\psi_{100}(r_1)\, dv = -\frac{e^2}{2a_H} S + \frac{e^2}{R} S - Q,$$

$$H_{22} = \int \psi_{100}^*(r_2)\, H\psi_{100}(r_2)\, dv = -\frac{e^2}{2a_H} + \frac{e^2}{R} - P.$$

Here use was made of the fact that ψ_{100} satisfies the Schrödinger equation of the H-atom with the eigenvalue $E_1 = -e^2/2a_H$ and the following simplifying notations have been introduced

$$P = \int \frac{e^2}{r_2}\, |\psi_{100}(r_1)|^2\, dv, \tag{66.6}$$

$$Q = \int \frac{e^2}{r_1}\, \psi_{100}^*(r_1)\, \psi_{100}(r_2)\, dv, \tag{66.7}$$

$$S = \int \psi_{100}^*(r_1)\, \psi_{100}(r_2)\, dv. \tag{66.8}$$

An interesting symmetry is seen in the matrix elements (66.5): $H_{11} = H_{22}$ and $H_{12} = H_{21}$. This is due to the fact that if the molecular ion is reflected with respect to the plane lying at half distance between the two nuclei and being perpendicular to direction R, i.e. if the transformation $r_1 \rightleftharpoons r_2$ is performed, the energy operator (Hamiltonian) (66.1) stays unchanged. Therefore if in the last integral of (66.5) the transformation $r_1 \rightleftharpoons r_2$ is performed, it changes into H_{11}, and similar relations hold for the quantities H_{12} and H_{21}. The same symmetry relations can be observed in the overlap matrix as well; the elements of this are, according to (64.19),

$$S_{11} = \int \psi_{100}^*(r_1)\, \psi_{100}(r_1)\, dv = 1\,,$$

$$S_{12} = \int \psi_{100}^*(r_1)\, \psi_{100}(r_2)\, dv = S\,,$$

$$S_{21} = \int \psi_{100}^*(r_2)\, \psi_{100}(r_1)\, dv = S\,,$$

$$S_{22} = \int \psi_{100}^*(r_2)\, \psi_{100}(r_2)\, dv = 1\,.$$

(66.9)

The energy E of the H_2^+ molecular ion can be found, using the model of (64.23) by solving the following quadratic equation:

$$\begin{vmatrix} -\dfrac{e^2}{2a_H} + \dfrac{e^2}{R} - P - E & -\dfrac{e^2}{2a_H}S + \dfrac{e^2}{R}S - Q - ES \\[2mm] -\dfrac{e^2}{2a_H}S + \dfrac{e^2}{R}S - Q - ES & -\dfrac{e^2}{2a_H} + \dfrac{e^2}{R} - P - E \end{vmatrix} = 0\,.$$

(66.10)

From this two E values are found:

$$E_s = -\frac{e^2}{2a_H} + \frac{e^2}{R} - \frac{P+Q}{1+S}\,,$$

(66.11)

$$E_a = -\frac{e^2}{2a_H} + \frac{e^2}{R} - \frac{P-Q}{1-S}\,.$$

(66.12)

For the first solution $c_1 = c_2$, for the second one $c_1 = -c_2$. That means that E_s is associated with a symmetric eigenfunction staying unchanged under the reflection $r_1 \rightleftharpoons r_2$. The eigenfunction, associated with E_a, is antisymmetrical; it changes sign under the reflection.

The integrals P, Q and S have still to be computed. It is the simplest to perform this by introducing the elliptic coordinates

$$\xi = \frac{r_1 + r_2}{R}\,, \qquad \eta = \frac{r_1 - r_2}{R}\,, \qquad \phi$$

(66.13)

where ϕ denotes the azimuthal angle denoting a rotation around the molecular axis joining the nuclei. These coordinates may take the following values:

$$1 \leq \xi < \infty, \quad -1 \leq \eta \leq 1, \quad 0 \leq \phi < 2\pi. \tag{66.14}$$

The expression for the volume element is

$$dv = \frac{1}{8} R^3(\xi^2 - \eta^2) \, d\xi \, d\eta \, d\phi. \tag{66.15}$$

The integrals P, Q and S reduce to elementary integrals if elliptic coordinates are used. For example, the expression for S is:

$$S = \int \psi_{100}^*(r_1) \, \psi_{100}(r_2) \, dv = \frac{1}{\pi a_H^3} \int e^{-(r_1+r_2)/a_H} \, dv$$

$$= \frac{\delta^3}{8\pi} \int_1^\infty d\xi e^{-\delta\xi} \int_{-1}^1 d\eta(\xi^2 - \eta^2) \int_0^{2\pi} d\phi = \frac{\delta^3}{2} \int_1^\infty d\xi \, \xi^2 e^{-\delta\xi} - \frac{\delta^3}{6} \int_1^\infty d\xi e^{-\delta\xi}$$

$$= \left(1 + \delta + \frac{1}{3} \delta^2\right) e^{-\delta}, \tag{66.16}$$

where the notation $\delta = R/a_H$ was used. Similarly, for P and Q we obtain without difficulty

$$P = \frac{e^2}{R} \left[1 - (1 + \delta) \, e^{-2\delta}\right], \tag{66.17}$$

and

$$Q = \frac{e^2}{a_H} (1 + \delta) \, e^{-\delta}. \tag{66.18}$$

Substituting these in the energy expressions E_s and E_a, the energy of H_2^+ is found as a function of the internuclear distance R:

$$E_s = \left[-\frac{1}{2} + \frac{1}{\delta} - \frac{1}{\delta} \cdot \frac{1 - (1 + \delta)e^{-2\delta} + \delta(1 + \delta) \, e^{-\delta}}{1 + (1 + \delta + \delta^2/3) \, e^{-\delta}}\right] \frac{e^2}{a_H}, \tag{66.19}$$

$$E_a = \left[-\frac{1}{2} + \frac{1}{\delta} - \frac{1}{\delta} \cdot \frac{1 - (1 + \delta)e^{-2\delta} - \delta(1 + \delta) \, e^{-\delta}}{1 - (1 + \delta + \delta^2/3) \, e^{-\delta}}\right] \frac{e^2}{a_H}. \tag{66.20}$$

These energy expressions are shown in Fig. 35. For very large R values, i.e. for $\delta \to \infty$, $E_s = E_a = -e^2/2a_H$. This is obvious as for very large separation the electron will form an unperturbed H-atom with one of the nuclei. For $R \to 0$ the energies tend to + infinity. It is seen that E_s lies always lower than E_a, i.e. the ground state will be the symmetric state. The minimum of E_s is found at an internuclear distance

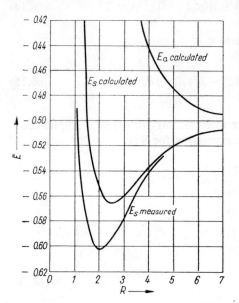

$$R_0 = 2.50\, a_H = 1.32 \times 10^{-8}\, \text{cm} \, ; \tag{66.21}$$

this value corresponds to the stable equilibrium position. The value of the energy is then

$$E_{s0} = -0.5646 \, \frac{e^2}{a_H} = -15.36 \, \text{eV} . \tag{66.22}$$

FIG. 35. The energy of the H_2^+ molecular ion as a function of the distance R from the nucleus. R is measured in a_H units, E in e^2/a_H units

There is no such stable position in the asymmetric state; with increasing R the energy decrease continuously.

The work needed to increase the internuclear distance R to infinity is called the dissociation energy of the molecule. In the dissociated state the energy, as already mentioned, will be equal to the energy $-e^2/2a_H$ of the H-atom. Thus the dissociation energy is

$$D = -\frac{e^2}{2a_H} - E_{s0} = 0.646 \, \frac{e^2}{a_H} = 1.76 \, \text{eV} . \tag{66.23}$$

The positive value of D shows that the molecule is stable against a spontaneous dissociation. This means that quantum mechanics can interpret theoretically the chemical binding of the H_2^+ molecular ion. The results obtained for the equilibrium internuclear distance and the dissociation energy D can be compared with the experimental data

$$R_0 = 1.06 \times 10^{-8} \, \text{cm} \quad \text{and} \quad D = 2.791 \, \text{eV} . \tag{66.24}$$

A better agreement cannot be expected from the very simple trial function (66.4).

It is noteworthy that the Schrödinger equation of H_2^+ can be solved exactly if parabolic coordinates are introduced. The exact solution yields the values $R_0 = 1.06 \times 10^{-8}$ cm and $D = 2.79$ eV, in full agreement with the experimental values (66.24).

The variational method enables the quantum-mechanical investigation of molecules being much more complicated than H_2^+. The interpretation of chemical binding and the theoretical deduction of the properties of molecules is one of the most beautiful results of quantum mechanics.

67. The Born Approximation

The Born approximation is the simplest method of approximation in the computation of scattering, it is used in the investigation of the scattering of high-energy particles. It is valid only if the bombarding energy E is much higher than the potential V of the scattering centre.

Let us start from the integral equation method of scattering calculus. The first step is to solve the integral equation (53.13), where a real k is substituted for q,

$$ u_l(r) = \frac{1}{k} \left[\frac{(2l+1)!!}{k^l} j_l(kr) + \int_0^r \{n_l(kr) j_l(kr') - j_l(kr) n(kr')\} U(r') u_l(r') dr' \right]. $$

$$(67.1)$$

If the potential U of the scattering centre is small, it is better to solve this equation by an iteration method. In zeroth approximation the scattering centre is totally neglected, i.e. U is taken equal to zero,

$$ u_l^{(0)}(r) = \frac{(2l+1)!!}{k^{l+1}} j_l(kr). $$

$$(67.2)$$

The next step of the iteration process is to substitute u_l in the second term containing U on the right-hand side expression of eq. (67.1). The first approximation thus is

$$ u_l^{(1)}(r) = \frac{(2l+1)!!}{k^{l+1}} \left[j_l(kr) + \frac{1}{k} \int_0^r \{n_l(kr) j_l(kr') - j_l(kr) n_l(kr')\} U(r') j_l(kr') dr' \right]. $$

$$(67.3)$$

223

The second approximation is gained by substituting this expression in the right-hand side of (67.1). The process may be continued until the required number of significant figures is obtained. After the computation of u_l the phase constant is obtained from (53.18).

In the following only the first Born approximation will be dealt with. In this case in (53.18) the second term of the denominator containing U is neglected, and in the numerator the zeroth approximation (expression (67.2)) will be used for u_l:

$$\delta_l = - \arctan \frac{1}{k} \int_0^\infty U(r) \, j_l^2(kr) \, dr .\qquad(67.4)$$

Substituting (48.3) and (50.3) the final value of the phase constant is

$$\delta_l = - \arctan \frac{4\pi^3 m_0}{h^2} \int_0^\infty V(r) \, J_{l+\frac{1}{2}}^2(kr) \, r \, dr .\qquad(67.5)$$

This means that in the Born approximation the computation of the phase constants is extremely simple. It is obtained from the potential $V(r)$ of the scattering centre by a single integration. Let us now compute the scattering amplitude (49.14). By using the identity

$$e^{2i\delta_l} = \frac{(1 + i \tan \delta_l)^2}{1 + \tan^2 \delta_l} = 1 + 2i \frac{\tan \delta_l + i \tan^2 \delta_l}{1 + \tan^2 \delta_l} ,$$

it can be written in the following form:

$$f(\vartheta) = \frac{1}{k} \sum_{l=0}^\infty (2l + 1) \frac{\tan \delta_l + i \tan^2 \delta_l}{1 + \tan^2 \delta_l} P_l(\cos \vartheta) .\qquad(67.6)$$

The Born formula has to be written into this expression. It should be taken into consideration that the Born approximation is valid only if the potential V and thus the phase constants δ_l are small, so that $\tan^2 \delta_l$ can already be neglected, and the result is

$$f(\vartheta) = - \frac{4\pi^3 m_0}{h^2 k} \int_0^\infty V(r) \left[\sum_{l=0}^\infty (2l + 1) J_{l+\frac{1}{2}}^2(kr) P_l(\cos \vartheta) \right] r \, dr .\qquad(67.7)$$

224

The sum in the square brackets is independent of the nature of the scattering centre. According the theory of Bessel functions this sum is equal to

$$\frac{\sin\left(2kr\sin\dfrac{\vartheta}{2}\right)}{\pi\sin\left(\dfrac{\vartheta}{2}\right)}.$$

Therefore the scattering amplitude in the Born approximation is

$$f(\vartheta) = -\frac{4\pi^2 m_0}{\hbar^2 k \sin\left(\dfrac{\vartheta}{2}\right)} \int_0^\infty V(r) \sin\left(2kr\sin\frac{\vartheta}{2}\right) r\, dr. \qquad (67.8)$$

It is an interesting property of this formula that it is a linear expression in the potential V. Therefore if the potential of the scattering centre is composed of two parts: $V = V_1 + V_2$, then the scattering amplitude will be the sum of the amplitudes f_1 and f_2, corresponding to V_1 and V_2, respectively: $f = f_1 + f_2$. From this linearity it also follows that increasing V n-times, the scattering amplitude f also increases n-fold. If the sign of V changes, also f changes sign. As the differential cross-section is equal to the square of the amplitude: $\sigma(\vartheta) = |f(\vartheta)|^2$, changing the sign of f does not change the cross-section. In the Born approximation the particles are scattered by a potential hill in the same way as by a potential well, being the image of the former.

As a simple application let us investigate the scattering by an exponential potential well:

$$V(r) = -V_0 e^{-r/a}. \qquad (67.9)$$

Substituting this in (67.8) an easily computable integral is reached. Introducing the notation

$$K = 2ka\sin\frac{\vartheta}{2}, \qquad (67.10)$$

the end result is

$$f(\vartheta) = \frac{8\pi^2 m_0 V_0 a}{\hbar^2 K} \int_0^\infty e^{-r/a}\sin\left(\frac{Kr}{a}\right) r\, dr = \frac{16\pi^2 m_0 V_0 a^3}{\hbar^2 (1 + K^2)^2}. \qquad (67.11)$$

Squaring this expression the differential scattering cross-section is found:

$$\sigma(\vartheta) = \left(\frac{16\pi^2 m_0 V_0 a^3}{\hbar^2}\right)^2 \cdot \frac{1}{\left(1 + 4k^2 a^2 \sin^2\dfrac{\vartheta}{2}\right)^4}. \qquad (67.12)$$

225

The largest cross-section, i.e. the strongest scattered intensity, is found for $\vartheta = 0$. This is the forward-scattered ray. The differential cross-section decreases with increasing ϑ. This decrease becomes stronger with increasing bombarding energy E, i.e. with increasing $k^2 a^2$. For very high energies only small angular deviations are obtained.

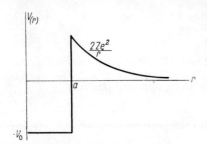

FIG. 36. The potential energy of the α-particles inside and in the vicinity of a nucleus

As a second example the scattering of α-particles by nuclei will be dealt with. It has been mentioned in Section 2 that for distances larger than the nuclear radius a the potential is $2Ze^2/r$, corresponding to Coulomb forces. For $r < a$, within the nucleus, the potential can be regarded as a spherical potential well (see Fig. 36).

Let us first regard the scattering by the potential well:

$$V_1(r) = -V_0, \quad \text{if } r < a,$$

$$\text{(67.13)}$$

$$V_1(r) = 0, \quad \text{if } r > a.$$

The corresponding scattering amplitude is received by substituting the potential of Fig. 36 in (67.8). Applying the abbreviation $K = 2ka \sin (\vartheta/2)$ the computation can be performed easily:

$$f_1(\vartheta) = \frac{8\pi^2 m_0 V_0 a}{h^2 K} \int_0^a \sin\left(\frac{Kr}{a}\right) r\, dr = \frac{8\pi^2 m_0 V_0 a^3}{h^2 K^3} (\sin K - K \cos K).$$

$$\text{(67.14)}$$

The other part of the potential is produced by the Coulomb field prevailing outside the nucleus:

$$V_2(r) = 0, \quad \text{if } r < a,$$

$$\text{(67.15)}$$

$$V_2(r) = \frac{2Ze^2}{r}, \quad \text{if } r > a.$$

The substitution of this in (67.8) yields the following integral:

$$f_2(\vartheta) = - \frac{16\pi^2 Z e^2 m_0 a}{h^2 K} \int_a^\infty \sin\left(\frac{Kr}{a}\right) dr . \qquad (67.16)$$

The value of this integral is indeterminate. It becomes determined by introducing a screening constant $e^{-\kappa r}$. (Such a screening is really produced by the space charge of the surrounding electrons.) The Born formula for the screened Coulomb potential is

$$f_2(\vartheta) = - \frac{16\pi^2 Z e^2 m_0 a}{h^2 K} \int_a^\infty e^{-\kappa r} \sin\left(\frac{Kr}{a}\right) dr$$

$$= - \frac{16\pi^2 Z e^2 m_0 a^2}{h^2 K} \frac{(\kappa a \sin K + K \cos K) e^{-\kappa a}}{\kappa^2 a^2 + K^2} .$$

Let us now investigate very weak screening, i.e. the limiting case of $\kappa \to 0$. Then the scattering amplitude is

$$f_2(\vartheta) = - \frac{16\pi^2 Z e^2 m_0 a^2}{h^2 K^2} \cos K . \qquad (67.17)$$

The potential shown in Fig. 36 is the sum of expressions (67.13) and (67.15). The amplitude of the α-scattering is thus received as the sum of (67.14) and (67.17):

$$f(\vartheta) = - \frac{16\pi^2 Z e^2 m_0 a^2}{h^2 K^2} \left[\cos K + \frac{V_0 a}{2 Z e^2} \left(\cos K - \frac{\sin K}{K} \right) \right] . \qquad (67.18)$$

The differential cross-section is

$$\sigma(\vartheta) = \frac{Z^2 e^4}{m_0^2 v_0^4 \sin^4\left(\frac{\vartheta}{2}\right)} \left[\cos K + \frac{V_0 a}{2 Z e^2} \left(\cos K - \frac{\sin K}{K} \right) \right]^2 , \qquad (67.19)$$

where $v_0 = hk/2\pi m_0$ is the initial velocity of the particles. The expression standing before the square brackets is exactly equal to (55.14), the cross-section of Coulomb scattering for a point-like nucleus. The expression

$$\left[\cos K + \frac{V_0 a}{2 Z e^2} \left(\cos K - \frac{\sin K}{K} \right) \right]^2 \qquad (67.20)$$

227

augmenting the Rutherford formula takes into consideration the correction due to the finite size of the nucleus which means that (67.19) accounts for the deviations from the Rutherford formula, i.e. for the anomaly of the scattering of α-particles. For very small nuclear radii, i.e. in the limiting case of $a \to 0$, the value of expression (67.20) becomes $+1$, i.e. the anomaly vanishes. The other limiting case is represented by a very high bombarding energy, i.e. an α-ray of very small wavelength. Then $K \gg 1$, and the value of the anomaly of (67.20) is

$$\left(1 + \frac{V_0 a}{2Ze^2}\right)^2 \cos^2 K \sim \frac{1}{2}\left(1 + \frac{V_0 a}{2Ze^2}\right)^2,$$

if $\cos^2 K$ is replaced by $\frac{1}{2}$, i.e. its mean value.

68. The Scattering of Electrons by Atoms

The scattering of high-energy (several hundreds of eV) electrons by atoms can be described by the first Born approximation. The electric field produced by the atom can be divided into two parts: one part originates from the nucleus, the other from the cloud of electrons. The resultant electric potential at a distance r from the nucleus is

$$\frac{Ze}{r} - e \int \frac{\rho(r')}{|\mathbf{r} - \mathbf{r}'|} \, dv' .$$

Here Z means the atomic number, $\rho(r)$ the density of the electron cloud, i.e. the number of electrons per unit volume. The potential energy of the bombarding electron in this field is

$$V(r) = -\frac{Ze^2}{r} + e^2 \int \frac{\rho(r')}{|\mathbf{r} - \mathbf{r}'|} \, dv' . \tag{68.1}$$

As the density distribution of the electron cloud $(\rho(r))$ is spherically symmetrical, the potential theoretical theorem (63.24) can be applied in solving this integral.

$$V(r) = -\frac{Ze^2}{r} + \frac{e^2}{r} \int_0^r \rho(r') \, 4\pi r'^2 dr' + e^2 \int_r^\infty \rho(r') \, 4\pi r' \, dr' . \tag{68.2}$$

228

Let us substitute this into expression (67.8) of the scattering amplitude. After introducing the notation $q = 2k \sin (\vartheta/2)$ we find that

$$f(\vartheta) = \frac{8\pi^2 m_0 e^2}{h^2 q} \left[Z \int_0^\infty dr \sin (qr) - 4\pi \int_0^\infty dr \int_0^r dr' \, r'^2 \rho(r') \sin (qr) \right.$$

$$\left. - 4\pi \int_0^\pi dr \int_r^\pi dr' \, r r' \rho(r') \sin (qr) \right].$$

Let us reverse the sequence of integration in the double integrals of the second and third term:

$$f(\vartheta) = \frac{8\pi^2 m_0 e^2}{h^2 q} \left[Z \int_0^\infty dr \sin (qr) - 4\pi \int_0^\infty dr' \int_{r'}^\infty dr \, r'^2 \, \rho(r') \sin (qr) \right.$$

$$\left. - 4\pi \int_0^\infty dr' \int_0^{r'} dr \, r \, r' \, \rho(r') \sin (qr) \right]. \qquad (68.3)$$

Now the integration with respect to r can be performed without the concrete knowledge of the density distribution $\rho(r')$. The integral of the last term over r is

$$\int_0^{r'} dr \, r \sin (qr) = \frac{r'}{q} \left(\frac{\sin (qr')}{qr'} - \cos (qr') \right). \qquad (68.4)$$

In the preceding section the undetermined integral of the second term has already been dealt with. After introducing the screening constant $e^{-\kappa r}$ and taking the limit as $\kappa \to 0$, the result is

$$\int_{r'}^\infty dr \sin (qr) = \lim_{\kappa \to 0} \int_{r'}^\infty dr \, e^{-\kappa r} \sin (qr) = \lim_{\kappa \to 0} \frac{[\kappa \sin (qr') + q \cos (qr')] e^{-\kappa r'}}{\kappa^2 + q^2}$$

$$= \frac{\cos (qr')}{q}. \qquad (68.5)$$

From this we obtain the first, single integral of (68.3) by the substitution $r' = 0$:

$$\int_0^\infty dr \sin (qr) = \frac{1}{q}. \qquad (68.6)$$

These results are substituted in the expression of the scattering amplitude and in the remaining integrals with respect to r', the integration variable can be denoted now by r:

$$f(\vartheta) = \frac{8\pi^2 m_0 e^2}{h^2 q^2}\left[Z - 4\pi \int_0^\infty \rho(r)\frac{\sin(qr)}{qr}r^2 dr\right]. \qquad (68.7)$$

This formula might be written also in the following form:

$$f(\vartheta) = \frac{e^2}{2m_0 v_0^2 \sin^2\left(\dfrac{\vartheta}{2}\right)}[Z - F(\vartheta)], \qquad (68.8)$$

where $v_0 = hk/2\pi m_0$ is the velocity of the bombarding electrons, and $F(\vartheta)$ is

$$F(\vartheta) = \frac{h}{m_0 v_0 \sin(\vartheta/2)}\int_0^\infty \rho(r)\sin\left(\frac{4\pi m_0 v_0 \sin(\vartheta/2)}{h}r\right)r\, dr. \qquad (68.9)$$

$F(\vartheta)$ is called the *form factor of the atom*. From (68.8) the differential scattering cross-section is

$$\sigma(\vartheta) = |f(\vartheta)|^2 = \sigma_R(\vartheta)\left[1 - \frac{F(\vartheta)}{Z}\right]^2, \qquad (68.10)$$

where

$$\sigma_R(\vartheta) = \frac{Z^2 e^2}{4m_0^2 v_0^4 \sin^4(\vartheta/2)} \qquad (68.11)$$

is the Rutherford cross-section, corresponding to (55.14). The form factor can be determined experimentally by measuring the cross-section $\sigma(\vartheta)$ of the electron scattering. Comparing this with (68.9) conclusions may be drawn about the distribution $\rho(r)$ of the electron cloud of the atom. The results obtained in that way are in good agreement with the theoretical density distributions determined by the quantum-mechanical eigenfunction of the atom.

Index

Adjoint spherical functions 111
Alkali atoms 31, 32
Alkali-earth metals 17, 32, 42
Alkali metals 17, 42
α-disintegration 170
α-particle(s) 6, 170 ff
  scattering of, by nuclei 6, 23, 226, 228
  velocity of 172
Amplitude 51
Angular momentum (moment of momen-
      tum) 14, 28, 36, 37, 72, 77, 112,
      114, 153
  eigenfunctions of 117, 153, 196
  operators assigned to the Cartesian co-
    ordinates of 112
Anharmonic 190
Anomalous Zeeman effect 198
Antiparallel spins 37
Atomic number, effective 216
Atomic weight 7
Azimuthal quantum number 28, 138

Balmer 16
  formula of 16
Band spectra 43
Basic spinor(s) 76, 115
Bergmann series 17
Bessel functions 123, 126, 155, 175, 225
β-rays 5
Binding energy 127
Bohr XI, 10, 18, 23, 24, 26, 27, 31, 33, 43,
     45, 48, 80, 84
  frequency condition 10, 24, 26, 80, 83
  magneton 12, 35, 36, 121
  orbits, radii of the spherical 139
  radius 195
  theory 90, 113, 138, 145
Bohr's quantum condition 24, 63, 64
Born 65
  approximation 223, 224, 228

Born — cont.
  formula 224, 227
Bracket series 17
Breit–Wigner formula 169

Canonically conjugate variables 47, 71,
  89
Cathode ray 3
Causality 50
Central potential field 108
Centre of gravity, coordinates of 146
Centre of mass, momentum of 147
Chadwick 6
Charge density 65
Chauchy XI
Chemical binding 42
Combining terms 33
Compton effect 20, 47, 48
Confluent hypergeometric series 181,
  182
Conjugate momenta 25
Continuity conditions 96, 99, 102, 105,
  123
Correspondence principle 32, 80
Coulomb forces 8, 226
  potential 143, 170, 177, 179, 227
  scatttering 179, 227
Critical potential 9, 10
Current 70
Current density 70

Dalton XI
Davisson 21
de Broglie 21, 53, 56, 57, 58
  wave 56, 57, 58
  wavelength 21, 58
Debye 21
Debye–Scherrer fringes 21
De Haas 13

16*

231

OTHER TITLES IN THE SERIES IN NATURAL PHILOSOPHY

238